Statistical Thermodynamics

With Applications to the Life Sciences

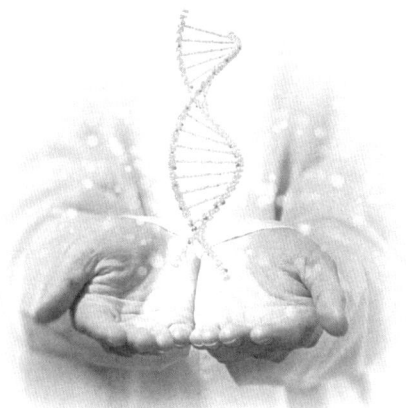

Statistical Thermodynamics
With Applications to the Life Sciences

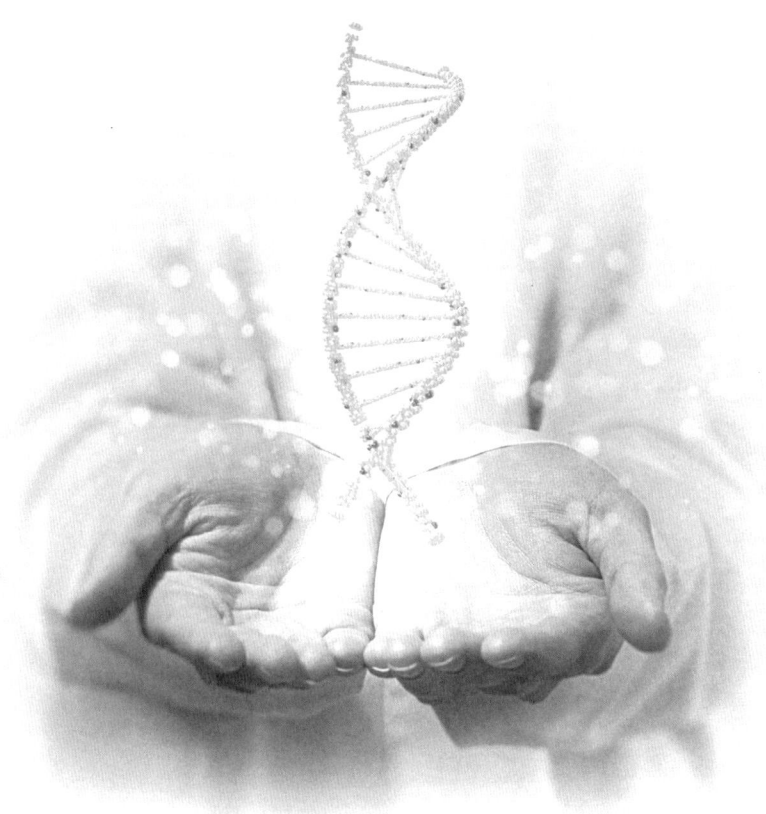

Arieh Ben-Naim

The Hebrew University of Jerusalem, Israel

 **World Scientific**

NEW JERSEY · LONDON · SINGAPORE · BEIJING · SHANGHAI · HONG KONG · TAIPEI · CHENNAI

Published by

World Scientific Publishing Co. Pte. Ltd.

5 Toh Tuck Link, Singapore 596224

USA office: 27 Warren Street, Suite 401-402, Hackensack, NJ 07601

UK office: 57 Shelton Street, Covent Garden, London WC2H 9HE

Library of Congress Cataloging-in-Publication Data
Ben-Naim, Arieh, 1934–   author.
    Statistical thermodynamics : with applications to the life sciences / Arieh Ben-Naim, The Hebrew University of Jerusalem, Israel.
        pages cm
    Includes bibliographical references and index.
    ISBN 978-9814579155 (hardcover : alk. paper) -- ISBN 978-9814578202 (pbk. : alk. paper)
    1. Biochemistry--Statistical methods. 2. Life sciences--Statistical methods. 3. Statistical thermodynamics. I. Title.
    QH323.5.B436 2014
    572.072'7--dc23
                                                                    2014007118

British Library Cataloguing-in-Publication Data
A catalogue record for this book is available from the British Library.

Typeset by Stallion Press
Email: enquiries@stallionpress.com

Printed in Singapore

Dedicated to Ruby

List of Abbreviations

eq.   equilibrium
HB   hydrogen bond
lhs   left hand side
1D   one dimensional
PF   partition function
PS   preferential solvation
rhs   right hand side
SMI   Shannon measure of information
SW   square well
ST   statistical thermodynamics
SI   symmetric ideal
3D   three dimensional
2D   two dimensional

Contents

Preface

This book is addressed to both students and researchers who are interested in *applying* statistical mechanics to understanding complex phenomena on a molecular level. Although this is not intended to be a textbook on statistical thermodynamics, as there is an abundance of these, it does contain a brief introduction to statistical thermodynamics, with the main purpose of bringing the reader to the level at which this powerful tool can be used in studying the problem of his or her interest. The reader is assumed to have some basic knowledge of thermodynamics.

As demonstrated in this book, one does not need to know much about statistical thermodynamics in order to apply it successfully. Only a very few "rules of the game" will suffice for that purpose. These rules are provided in the "Introduction" of the book. Once the reader *adopts* these rules even on a very rudimentary level, the door opens to an immense landscape of possible applications; some of which are discussed in this book while others, more importantly, can be devised and studied by the reader. The experience in working with simple models can be rewarding in two senses. First, one gains confidence in the usefulness of statistical thermodynamics. Second, from studying simple models one can learn a great deal about the molecular "mechanism" underlying seemingly very complex phenomena involving interacting particles. The rest of the book consists of various applications, from the simplest Ising model to the highly complex systems such as aqueous solutions and allosteric phenomena.

Statistical thermodynamics is an extremely useful and powerful tool for studying complex systems by using "minimal molecular models," i.e. models that contain the minimal number of features that produce a given phenomenon. For instance, in order to understand why liquid water contracts between 0°C to 4°C, an almost unique property of water, it is not necessary to study a model of water molecules in three dimensions. An extremely simplified 1D model is sufficient, whereas real water molecules in 3D interact in a very complex way via a six-dimensional function. The water-like particles in 1D interact via a very simple one-dimensional function. The art of the game is to find out what the *essential cause* of a phenomena is, and then implement it in a "minimal model," i.e. a model which contains only the essential features that are needed in exhibiting the studied phenomena.

Of course, it is not always easy to find out what the essential features of the interactions are between particles that lead to some macroscopic behavior of a real system. It is also not always clear how to implement them in a 1D model even after having identified the essential features. However, once these are found the study of these models becomes very satisfying.

To return to the example of water, it has long been known that the "structure of water" is a key feature of the molecular properties of water, and that the structural breakdown upon heating gives rise to the negative temperature dependence of the volume of water at low temperatures. If the "structure of water," as it is currently understood is the key molecular feature of water, then one cannot hope to implement this feature in either a 2D or 1D model. Nevertheless, from numerous studies of water and aqueous solutions, it was found that the "structure of water" per se, is not essential for the understanding of the unique properties of water. It was found that a more fundamental principle underlies both the structure of water, as well as the unique behavior of this liquid. This principle is described in Chapter 10.

Similarly, the efficient transportation of oxygen by hemoglobin and the mechanism of regulation of biochemical processes by regulatory enzymes are highly complicated phenomena. Nevertheless, there is a fundamental principle underlying both of these phenomena. When this principle is implemented in a simple model one can learn a great deal about the molecular events that produce these phenomena.

The book is designed in such a way that it will provide the reader with some general ideas of the type of problems for which statistical thermodynamics can be useful. Most of these "simple models" are chosen as one-dimensional (1D).

There are several reasons for choosing 1D models (whenever possible).

(1) There are many systems that are actually 1D system such as linear polymers, proteins, nucleic acids, etc.
(2) Most of these models can be solved exactly. This is a great advantage over models that cannot be solved exactly.
(3) The mathematical techniques of solving 1D models are elegant and useful in many other fields.

Bear in mind that many phenomena such as cooperativity and regulation by allosteric enzymes, etc. are not 1D system. However, one can still design a 1D model to study such a system. All that one needs is that the patterns of interactions between the 1D elements (or subunits) be in one dimensional, and that the essential features of the phenomena be incorporated in the model.

Some sections of this book contain details of mathematical derivation. These are marked by triple asterisks (***). If you are interested only in getting a qualitative idea about the model used and the results, you can skip the mathematical parts on first reading and proceed to the results. On the other hand, if you want to apply the model, or a related model to your own problem, then it will be necessary to follow the details of the mathematical procedures.

I believe that this way of designing the book will eventually encourage the reader to apply the methods of statistical thermodynamics to his/her own problems even if they seem to be very complex at first sight. I also believe that they will find the mathematical methods elegant and beautiful, and perhaps overcome their fear of mathematics.

Acknowledgments

I would like to thank my friends and colleagues Frank Bierbrauer, Diego Casadei, Richard Henchman, Robert Hanlon, Joaquim Mendes and Samuele Zampini for reading parts or all of the manuscript, detecting errors and mistakes, and offering some useful comments. I would also like to express my appreciation and thanks to my wife Ruby for her help in typing, editing and re-editing the book patiently.

Part I

Fundamentals

1

Introduction

1.1 What is Statistical Thermodynamics?

Statistical thermodynamics (ST), sometimes also referred to as *statistical mechanics*, is a very powerful mathematical tool for the study of the *thermodynamic* properties of matter at equilibrium, from the knowledge of the properties of the individual molecules comprising the macroscopic system. Statistical thermodynamics may be viewed as a bridge between the microscopic world of atoms and molecules, on the one hand, and the macroscopic world as we observe it in our daily lives, on the other hand.

These two realms of matter are quite different. Most of our knowledge about the properties of atoms and molecules comes from indirect inference, such as from the measurement of the dielectric constant, deviation from the behavior of ideal gases (virial coefficients), spectra, nuclear magnetic resonance (NMR), etc. On the other hand, the macroscopic properties of matter such as heat capacity, heat of evaporation, etc. are directly measurable. Statistical thermodynamics, as the term implies applies statistical methods to a large number of *microscopic* particles to obtain *averages*. These averages are the thermodynamic quantities, i.e. the macroscopic properties of matter. But these methods provide more than just numerical values of thermodynamic quantities. They can also contribute to our *understanding* of many outstanding phenomena. It is this aspect of statistical thermodynamics that we shall emphasize and pursue in this book. Thus, instead of trying to calculate, say, the heat capacity of liquid water, which is a very important quantity, we shall be interested in the molecular origin of this quantity. Instead of calculating the binding curve (or the binding isotherm) of oxygen to hemoglobin, we shall be interested in the origin of the *cooperativity* of the binding, which gives rise to the particular shape of the binding curve.

1.2 What are the Differences Between Statistical Thermodynamics and Thermodynamics?

Thermodynamics and statistical thermodynamics deal with the same quantities: the thermodynamic quantities of a macroscopic system at equilibrium, e.g. energy, pressure, temperature, etc. However, there are some fundamental differences in the methodology of studying these thermo-dynamic properties. Thermodynamics deals exclusively with macroscopic quantities and their relationships. It does not *explain* how these properties arise from the properties of atoms and molecules. Although it is recognized that the system is composed of atoms and molecules, thermodynamics does not make use of the atomic constituency of matter.

Thermodynamics provides general relationships between various thermodynamic quantities, e.g.

$$\left(\frac{\partial S}{\partial P}\right)_{T,N} = -\left(\frac{\partial V}{\partial T}\right)_{P,N} \qquad (1.1)$$

This is not a trivial relationship. It states that the pressure dependence of the entropy (at constant T and N) is related to the temperature dependence of the volume (at constant P and N) of the system. This is not trivial, and I guess you would not be able to guess it unless you learn thermodynamics.

Another example is the *Clausius equation*

$$\left(\frac{\partial P}{\partial T}\right)_{eq} = \frac{\Delta H}{T \Delta V} \approx \frac{\Delta H}{RT^2} \qquad (1.2)$$

which gives the temperature dependence of the vapor pressure of a pure liquid at equilibrium (eq), in terms of the heat of vaporization (ΔH), and the volume change upon vaporization (ΔV). Note that the first equality is exact and valid for any substance. The second is an approximate one, and is known as the *Clausius-Clapeyron equation*.

All these relationships are said to be *universal*; they apply to any substance. Thermodynamics does not aim at calculating the *values* of any of these quantities. On the other hand, statistical thermodynamics is in principle, designed for calculating the values of say, the entropy, the energy, etc. of a specific system. I wrote "in principle," because in practice, there are very few systems for which we can really calculate the *values* of their thermodynamic quantities.

When we discuss thermodynamic quantities we usually do not explicitly write the functional relationship, say between entropy, temperature and pressure. On the other hand, in statistical thermodynamics we deal with the *explicit functions*.

For a classical monatomic ideal gas we have the explicit function for the entropy

$$S(T, V, N; m) = N k_B \ln \left[\left(\frac{2\pi m k_B T}{h^2} \right)^{3/2} \frac{e^{5/2} V}{N} \right] \quad (1.3)$$

Note that this formula provides an *explicit* dependence of the entropy S on the macroscopic parameters T, V, N as well as on the microscopic property m, the molecular mass of the atoms. For each gas, we have a *different function* depending on the parameter m. At this stage the function in Eq. (1.3) is presented as an example. We shall discuss its derivation and its meaning in the following chapters.

As it stands, Eq. (1.3) for the entropy looks complicated and perhaps intimidating. You might ask yourself: Why should I study this expression? Why should I memorize this complicated formula? What information does it convey? My answer is that you should not memorize this equation. Later we will have an in-depth discussion about this equation. I brought it up here as an example, a relatively rare example in statistical thermodynamics of an *explicit* expression of the entropy of a *specific* system. The purpose is to compare the way we handle the entropy in thermodynamics and in statistical thermodynamics. In Chapter 3 we shall see that not only does Eq. (1.3) provide a *value* for the entropy of an ideal gas that agrees with the experimental value, an extraordinary achievement of statistical thermodynamics, but it also provides a *meaning* to entropy, which is not a less important achievement. Both these achievements would not be possible in the realm of thermodynamics. Thermodynamics, as we noted above, presumes that there exists a quantity, which Clausius called the entropy, which has some properties and fulfills some relationships with other thermodynamic quantities. Thermodynamics never provides a molecular interpretation of a thermodynamic quantity. This is the reason why, for a very long time, entropy has been considered as one of the most mysterious quantities in Physics.[1] I hope that when you get to Chapter 3, you will see that there is no mystery involved in the concept of entropy.

Let us continue comparing thermodynamics and statistical thermodynamics. In thermodynamics we usually do not bother to specify the *independent variables* we choose to describe the system. For instance, a one-component system may be described by two intensive variables, say (P, T), (μ, T), (ρ, T), etc. Two intensive variables are sufficient to describe the thermodynamic properties of the system. If you are interested in the *size* of the system you can also add one extensive quantity, say, the volume or the total number of particles in the system.

Any choice of independent variables is acceptable. Thus, when we discuss the relationship (Eq. 1.1) we assume implicitly that the entropy $S(T, P, N)$ is a function of the independent variables, and we take the partial derivative of this function with respect to P, keeping T and N constants. We could have also written a derivative of the form $(\partial S / \partial \rho)_{T,N}$ which implicitly means that we choose to view the entropy as a function of the variables T, ρ, N, and we take the partial derivative of the function $S(T, \rho, N)$ with respect to the density ρ, keeping T and N constant.

In thermodynamics we are at liberty to choose any set of independent variables, in terms of which we can write the different functions: $S(T, P, N), S(T, \rho, N), S(T, V, \mu)$, etc. These are *different* functions but in thermodynamics they are considered to be equivalent. In statistical thermodynamics we can also study the different functions $S(T, V, N)$, $S(T, P, N)$, etc. but these are not equivalent. For instance, the function $S(U, V, N)$ where U is the internal energy of the system, is a more fundamental function than, say $S(T, V, N)$, in the sense that one can derive all the thermodynamic quantities of interest from the former. Thus, we shall refer to $S(U, V, N)$ as the *fundamental function* of an isolated system (i.e. a system with constant energy, volume and number of particles or number of moles of particles). Other fundamental functions that we shall use are Gibbs energy $G(T, P, N)$, and the Helmholtz energy $A(T, V, N)$, etc. We shall say that the *fundamental function* for a system described by the independent variables T, P, N is the function $G(T, P, N)$. Likewise, the *fundamental* function for a system described by the independent variables T, V, N is the function $A(T, V, N)$.

The final difference between statistical thermodynamics and thermodynamics that we shall discuss here is the presumed *sharp* vis-à-vis the actual *fluctuating* values of the thermodynamic quantities. In thermodynamics we always assume that each (macroscopic) thermodynamic quantity has a well defined value, i.e. a sharp value. In statistical thermodynamics we also recognize fluctuations in the values of the thermodynamic quantities. For example, a one-component system may be described by the independent variables T, P, N. If we fix the temperature T, the pressure P, and the number of particles (or the number of moles), such a system in thermodynamics has a definite (sharp) volume V. For instance, for an ideal gas the volume is given by

$$V = \frac{Nk_B T}{P} \tag{1.4}$$

where k_B is the Boltzmann constant. For a general system we do not have such a neat *equation of state* that allows us to calculate the volume of a system given the values of P, T, N. Nevertheless, we assume that if the system is at equilibrium the volume is *determined* by the independent variables T, P, N, i.e. there *exists* a function $V(T, P, N)$, which we write as

$$V = V(T, P, N) \tag{1.5}$$

Note that the letter V is used here for the *value* of the volume, as well as for the *name* of the *function*, relating the volume of the system to the independent variables T, P, N.

In statistical thermodynamics the same system described above can have a *distribution* of volumes, not a sharp value V. We talk about the probability $[Pr(V)]$ of finding the system with some volume V. This distribution function is very narrow for a macroscopic system, as shown in Fig. 1.1. The *average* value of the volume is given by

$$\langle V \rangle = \int_0^\infty V Pr(V) dV \tag{1.6}$$

Thus, in principle any volume between zero to infinity is possible. In practice only a very narrow range of values of

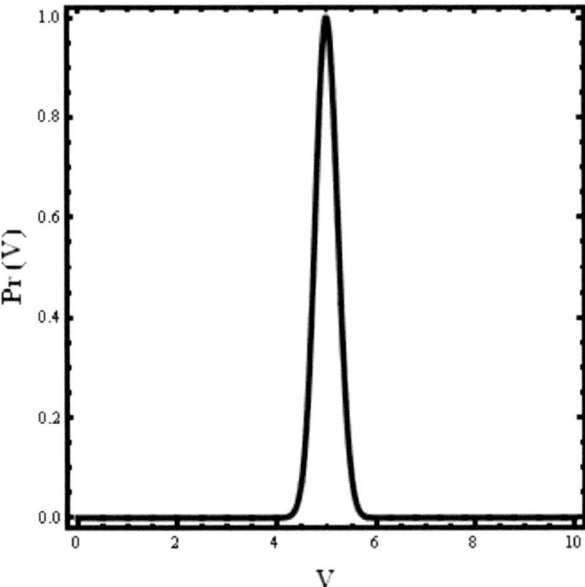

Fig. 1.1 A sharp distribution of the probability density $Pr(V)$. Note that when the distribution density is very sharp one cannot distinguish between the average volume (Eq. 1.6), and the most probable volume of the system.

the volume have a significant probability. All other values of the volumes outside this range have very low probabilities. We shall discuss a measure of the *size* of the fluctuations, i.e. the width of the distribution function $Pr(V)$, in Section 1.5. Note that whenever the distribution is very sharp, we do not distinguish between the average quantity $\langle V \rangle$, and the most probable quantity, i.e. the volume V for which $\frac{\partial Pr(V)}{\partial V} = 0$.

1.3 The Basic Postulates of Statistical Thermodynamics

There are essentially two postulates upon which the whole edifice of statistical thermodynamics is erected. The atomic constituency of matter was once regarded as a third postulate. Today, this is no longer considered a postulate, but a fact.

Postulates, like axioms, are statements that everyone, or at least those who want to use statistical thermodynamics, has to agree on. The postulates are not obviously true, as so many axioms that you might have encountered in geometry or in number theory. However, I shall try to present these postulates so that they *make sense*, and therefore are easy to accept. In fact, you do not need to know much about the postulates, or how the theory of statistical thermodynamics is built upon the postulates, if you are only interested in applying the theory to solving your own problem.

The first postulate

The time average measured on a single macroscopic system is equal to the ensemble average.

What this means is very simple to understand. It is often referred to as the *Ergodic* postulate.

Consider a macroscopic system of, say, a one-component gas at a fixed pressure P and temperature T. We measure the volume of the system and denote the result by V. If the system is macroscopically large we shall find the same value V any time we measure the volume of the system. We must recognize that each measurement takes some finite interval of time, say $\Delta t \approx 1$ sec, and what we actually measure is not the *value* of the volume at a fixed "point-time," but an average value of the volume — averaged over the time interval Δt. Let us denote this time average (*ta*) by the value $\langle V \rangle_{ta}$. Remember that within the time of measurement Δt, many molecular events have occurred, and the average volume we have measured is a time average over all these molecular events, each of which might be associated with a slightly different volume V. Unfortunately, we cannot calculate the exact volume of the system at each point-time t, to compute the function $V(t)$, and then average over the time interval Δt. Here comes the brilliant idea of Gibbs. Suppose that instead of measuring the volume of a single system at each point-time t, we envisage an *ensemble* of systems. The ensemble contains a very large number of systems M, each of which is characterized by the same independent variables T, P, N. Each system in the ensemble has, by definition, the same macroscopic values of T, P, N. If we measured the volume of each system in the ensemble at one point-of-time t, we would have gotten many different values of the volumes; V_i for the ith system in the ensemble.

The first postulate states that the time average of the volume of a single system will be equal to the *ensemble* average, provided that all the systems in the ensemble are characterized by the same thermodynamical variables T, P, N.

This is a plausible postulate. If you have any qualms about its plausibility, consider the following simple example. Suppose you throw a dice many times. You "measure" the outcomes, which are numbers between one and six. If you repeat these measurements many times you will get a series of results $n(1)$, $n(2)$, $n(3)$, . . . You can also calculate the average result, which is the analog of the time average. Instead of calculating the time average, imagine that you throw a very large number of dice, all dice being identical. You can look at all the results of all the dice in the ensemble and calculate the average result over the ensemble. This will be the analog of the ensemble average.

If you actually do this experiment and measure the average result over many throws of a single dice, or the average over the ensemble of dice, you might get slightly different results say, 3.51 and 3.49 for an "ideal" dice. We believe however, that if we repeat the same experiment many, many times, and measure the average over numerous throws of dice in the ensemble we should obtain the same result.

Before going on to discuss the second postulate it should be noted that the analogy between the thermodynamic system and the dice is not perfect. We have implicitly assumed that the measurement in the series of throws was made with the same dice, and that the results are statistically independent. In reality, if you throw the *same* dice many times it might slightly change after each throw, for instance its shape might be deformed or part of the dice might chip off. Therefore, we cannot ascertain that the condition of the dice at the millionth throw is exactly the *same* as in the first throw. In other words, the events might be dependent; the result at the kth throw might affect the probability of the result of the $k + 1$th throw.

Similarly, when we observe in our mind, the evolution of the state of a single system, we cannot claim that the results obtained at different times are independent. In fact the state of the system at $t + \Delta t$ is very much dependent on the state of the system at time t. Nevertheless, we adopt this postulate, not because it is "obviously true" as an axiom would be, but because the theory based on this postulate is successful, in the sense that it leads to prediction of macroscopic quantities, which conform with the experimentally measured quantities.

The second postulate

In an isolated system all possible microstates are equally probable.

Again, this is a plausible postulate. It is not "obviously correct" as an axiom is, and it cannot be proved correct. However, it is plausible for the following reason.

Let us again use the analogy between throwing the dice and the systems in the ensemble. If the dice are *known* to be fair, it is equivalent to saying that each of the six outcomes has the same probability. However, if we do not know anything about the dice, the *best guess* we can make is to assume that the dice are fair and therefore each outcome has the same probability of occurrence.

This argument can be rephrased using the Shannon measure of information, which we shall discuss in the next chapter.

As in the case of the first postulate, the analogy between a dice and a system in an ensemble is not perfect. All we can say is that a macroscopic isolated system (i.e. having a fixed energy, volume and number of particles) has many quantum mechanical states (in classical systems the characterization of the state of the system uses the continuous locations and momenta of all the particles). Having no other information on the relative likelihood of these states we can venture a guess and our best bet is that each state is as likely to occur as any other state.

On these two postulates and with some mathematical procedures the whole theory of statistical thermodynamics was constructed. If you are interested in applying the theory for solving your own problems, you need not spend time and effort to studying the details of the theory, in the same way that you need not spend time and effort on the details of electronics in order to make use of the capabilities of your computer.

1.4 Ensembles, and the Structure of Statistical Thermodynamics

In this section we present the main "rules of the game." We shall not explain how these rules were derived. We shall simply adopt them, much as we accept the rules of chess or the rules of any other game. Once you adopt the rules, you can play the game. As in any game, the theory of statistical thermodynamics should be entertaining. But the theory of statistical thermodynamics has more to offer than mere entertainment. It is also an extremely useful tool to penetrate into the microscopic world and "see" the molecular reasons underlining some observable phenomena.

In the following sections we shall present the fundamental equations, and the rules of how to use them according to the different ensembles, or the different set of independent variables we choose to characterize the thermodynamic systems. In Part II, we shall apply these rules to study specific systems and phenomena.

1.4.1 *The Isolated System; the E, V, N Ensemble*

An isolated system is defined as a system that has a fixed energy E, a fixed volume V, and a fixed number of particles N (if the system contains several components, N is replaced by the vector $N = (N_1, N_2, \ldots, N_c)$ where N_i is the number of particles of the ith kind).

A truly isolated system does not exist. A real system is never perfectly isolated from its surroundings. Even if such a system existed, it would be of no interest to us. We could not make any measurements on it, nor observe its behavior. Any measurement, or observation entails *interaction* with the system, and that is, by definition of the isolated system, impossible. Nevertheless, the isolated system is a convenient starting point for building the theory.

Let Ω be the number of quantum mechanical states of the isolated system characterized by the variables E, V, N. You do not need to know what these states are, or how one can calculate them. All you need to know is that a macroscopic system has a huge number of (Ω) microscopic states, something of the order of N^N. If N is of the order of 10^{23}, which is huge, then N^N is unimaginably larger than N.

If you feel uncomfortable with the *quantum mechanical* state, think of a classical system. Each particle is characterized by its location and momentum (or velocity). A *state* of such a system is a list of all the locations and all the momenta of all the N particles; this is a huge vector $(R_1, R_2, \ldots, R_N, p_1, p_2, \ldots, p_N)$ where each of the R_i and each of the p_i has three components (x_i, y_i, z_i) and (p_{ix}, p_{iy}, p_{iz}). Altogether, this vector has 6N components. Can you imagine how many states of such a system are consistent with the macroscopic characterization E, V, N? Do not try to count them; there are an infinite number of states, presuming that this system behaves *classically*. The reason is that each of the variables R_i and p_i can change continuously.

In quantum mechanics the *state* of the system is described differently. Although it sounds more abstract than the description in terms of locations and momenta, the usage of the quantum mechanical states is easier in formulating the fundamental equation for the isolated system. Assuming that there are Ω states, we define the *Boltzmann entropy* by

$$S_B = k_B \ln \Omega \tag{1.7}$$

where $k_B = 1.38 \times 10^{23} \, \mathrm{JK}^{-1}$ is called the *Boltzmann constant*. Note carefully that S_B is *defined* in Eq. (1.7). At this moment I urge you to accept on faith that S_B, as defined in Eq. (1.7) behaves, up to an additive constant, exactly as the thermodynamic entropy of an isolated system. This means that all the thermodynamic quantities derived from this equation behave the same as the thermodynamic quantities you have encountered in thermodynamics. Therefore, we shall drop the subscript B in S_B. In Chapter 3, we shall also see that the entropy of an ideal gas may be obtained from Shannon's measure of information. This will provide you with a new and profound meaning of the thermodynamic entropy.

Accepting the identity of S_B with the entropy of the system, we take the total differential of the function $S(E, V, N)$, and identify the partial derivatives with the partial derivatives of the thermodynamic entropy, i.e.

$$dS = \left(\frac{\partial S}{\partial E}\right)_{V,N} dE + \left(\frac{\partial S}{\partial V}\right)_{E,N} dV + \left(\frac{\partial S}{\partial N}\right)_{E,V} dN$$

$$= \frac{1}{T} dE + \frac{P}{T} dV - \frac{\mu}{T} dN \tag{1.8}$$

where T is the absolute temperature, P is the pressure, and μ is the chemical potential (if there are c-components, then μ is the vector $\boldsymbol{\mu} = (\mu_1, \mu_2, \ldots, \mu_c)$).

Finally, we introduce the fundamental distribution of the states in the isolated system. Following the second postulate of statistical thermodynamics we assert that if there are Ω states, and if they are equally probable, then the distribution $(\text{Pr}(1), \ldots, \text{Pr}(\Omega))$ must satisfy the equation

$$1 = \sum_{i=1}^{\Omega} \text{Pr}(i) = \sum_{i=1}^{\Omega} p = p\Omega \qquad (1.9)$$

Therefore,

$$\text{Pr}(i) = p = \frac{1}{\Omega} \qquad (1.10)$$

Exercise: Show that the entropy, as defined in Eq. (1.7), may also be written as

$$S = -k_B \sum_{i=1}^{\Omega} Pr(i) \ln Pr(i)$$

where $Pr(i)$ are given in Eq. (1.10). The significance of this representation of the entropy will be discussed in Chapter 2.

1.4.2 The Isothermal System; the T, V, N Ensemble

In the previous subsection we constructed an ensemble of M isolated systems, each of which is characterized by the variables (E, V, N). We also assumed that the systems in the ensemble do not interact with each other. In this and in the following subsection we shall relax the constraint on "no interactions," and allow different types of interactions among the systems in the ensemble.

First, suppose that the thermally insulated boundaries (or *athermal*) in the previous ensemble (Fig. 1.2) are replaced by thermally conducting (or *diathermal*) boundaries (Fig. 1.3). The ensemble as a whole is still considered to be isolated. We allow the *flow* of *heat* between the individual systems in the ensemble, while still keeping the volume V and the number of particles in each system fixed.

What would happen?

Clearly, when the athermal walls are replaced by heat conducting walls, heat will flow between the systems, and if we take a "snapshot," and measure the energy of each system

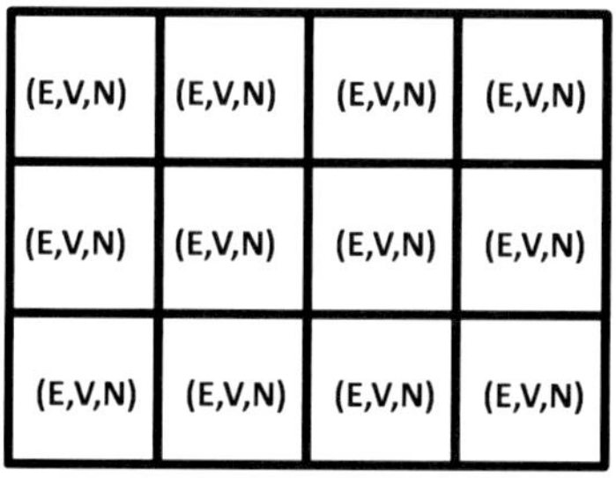

(E,V,N)	(E,V,N)	(E,V,N)	(E,V,N)
(E,V,N)	(E,V,N)	(E,V,N)	(E,V,N)
(E,V,N)	(E,V,N)	(E,V,N)	(E,V,N)

Fig. 1.2 An ensemble of isolated systems, each having the same (E, V, N).

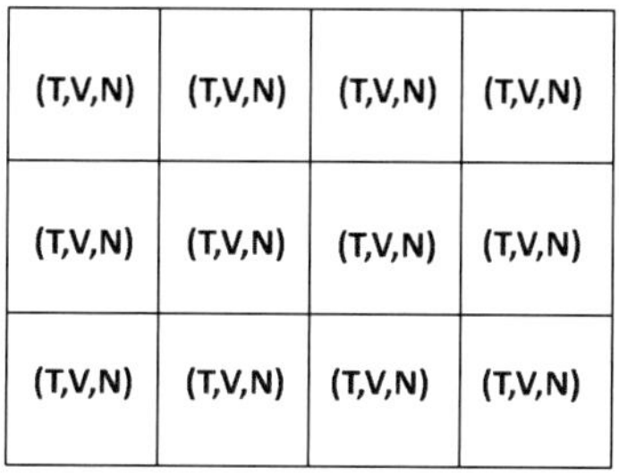

(T,V,N)	(T,V,N)	(T,V,N)	(T,V,N)
(T,V,N)	(T,V,N)	(T,V,N)	(T,V,N)
(T,V,N)	(T,V,N)	(T,V,N)	(T,V,N)

Fig. 1.3 A (T, V, N) ensemble is obtained by replacing the athermal boundaries by diathermal boundaries so that heat may flow from one system to the other.

in the ensemble, we shall find that the value of E is different for each system in the ensemble.

In order to calculate average quantities in this ensemble one does not need to introduce a new postulate. Instead, one implements the second postulate on the entire ensemble and then derives the required probabilities, which in our case are

$$Pr(E) = \Omega(E, V, N) \exp[-\beta E]/Q \qquad (1.11)$$

where β is used as a short hand notation for $(k_B T)^{-1}$, and $Pr(E)$ is the probability of finding a system in the ensemble having energy E. $\Omega(E, V, N)$, as before is the *degeneracy*, i.e. the number of states having the same energy E (and V, N as well), and Q is a normalization constant, which is determined by the requirement

$$\sum_E Pr(E) = 1 \qquad (1.12)$$

where the sum is over all possible energy "levels" of the system.

From Eq. (1.11) and (1.12) we obtain

$$Q(T, V, N) = \sum_E \Omega(E, V, N) \exp[-\beta E] \qquad (1.13)$$

The quantity $Q(T, V, N)$ is referred to as the T, V, N *partition function* (PF).

Before we move on, note carefully what we have achieved so far. We started with an ensemble of isolated systems, each characterized by the same values of the macroscopic variables E, V, N. The systems in the ensemble are "identical" from the *macroscopic* point of view. However, viewed microscopically they might be in different microscopic states.

Once we open the boundaries to heat flow, the parameter E of each system in the ensemble is no longer fixed (but V and N remain unchanged). Instead there will be a *distribution* of energies. In other words we can talk about the probability of finding a single system having an energy E, or equivalently the *fractions* of *systems* in the ensemble having some energy E (here we have assumed that the possible energies are discrete, say $E_0, E_1, E_2 \cdots$ and not continuous). We claimed that this probability distribution is given by Eq. (1.11). The important thing to remember is that this probability distribution is *derived* from the second postulate. The details of the derivation are discussed in any textbook on statistical thermodynamics.

Accepting the distribution in Eq. (1.11) we next imposed the condition on Eq. (1.12), which simply means that the probability of finding a system in *any* energy level is one. This procedure gives the normalization constant Q. Note that in Eq. (1.13) we wrote $Q(T, V, N)$, and not just Q. Why?

We started with the function $\Omega(E, V, N)$, i.e. the number of states of a system characterized by the variables E, V, N. We then multiplied it by the quantity $\exp[-\beta E]$ and sum over all values of E. Once we had summed over all E the result was no longer a function of the variable E. Instead, we introduced a new parameter β, which is $\beta = (k_B T)^{-1}$, where k_B is the Boltzmann constant, and T the absolute temperature. The resulting quantity is now a function of T instead of E (V and N remain unchanged).

The physical reason underlying this change in the thermodynamic variables is as follows. We know from thermodynamics that if we take two isolated systems and bring them to thermal equilibrium, heat will flow from the hot to the cold body. The energy of each system will no longer be fixed, but the two systems will be characterized by a fixed temperature. This is of course true for any number of systems. In our case we started with an ensemble of M systems, each characterized by the same energy E (as well as V and N). When we bring them to thermal contact, heat will be exchanged between the systems. The energy of each system will not be fixed, but will fluctuate (we can actually calculate the amount of these fluctuations, see Section 1.5). However, the temperature of the entire ensemble will be constant.

The formal transformation of variables from E to T (or equivalently from E, V, N to T, V, N) is achieved in the summation (Eq. 1.13), which is the discrete analog of the Laplace transform.

We shall now make the connection between the new quantity $Q(T, V, N)$, and the Helmholtz energy $A(T, V, N)$. This connection is very important.

$$A(T, V, N) = -k_B T \ln Q(T, V, N) \qquad (1.14)$$

We shall refer to Eq. (1.14) as the *fundamental equation* of the T, V, N ensemble (sometimes also referred to as the *canonical ensemble*). We shall present here a quick argument to show how this relationship comes about. Starting from the definition of Q in Eq. (1.13) and using the fundamental equation of the E, V, N ensemble, we have

$$Q(T, V, N) = \sum_E \exp[S(E, V, N)/k_B] \exp[-\beta E]$$

$$= \sum_E \exp\{-\beta[E - TS(E, V, N)]\}$$

$$= \sum_E \exp[-\beta A(T, V, N; E)]$$

$$\approx \exp[-\beta A(T, V, N)] \qquad (1.15)$$

Note that in Eq. (1.15) we identified the difference *E-TS* as the Helmholtz energy. There is one *approximation* which is made in the last step; replacing a sum over many positive terms by one term only; the maximal term. This is an extraordinary approximation, which can be shown to be justified in the limit of macroscopic systems.[2] It should also be noted that E is used both as the energy of the isolated system, and as an independent variable, in Eq. (1.15).

Once we have the function $Q(T, V, N)$, then together with the fundamental Eq. (1.14) we can derive all the thermodynamic quantities of interest. To do this we can use the total differential of the Helmholtz energy (expressed as a function of the variables T, V, N)

$$dA = -SdT - PdV + \mu dN \qquad (1.16)$$

from which we can obtain all the thermodynamic quantities, e.g.

$$S = \frac{-\partial A}{\partial T} = k_B \ln Q + k_B T \frac{\partial \ln Q}{\partial T} \qquad (1.17)$$

$$P = \frac{-\partial A}{\partial V} = k_B T \frac{\partial \ln Q}{\partial V} \qquad (1.18)$$

$$\mu = \frac{\partial A}{\partial N} = -k_B T \frac{\partial \ln Q}{\partial N} \qquad (1.19)$$

Exercise: Show that the entropy of the system in the (T, V, N) ensemble may be written as

$$S = -k_B \sum_E Pr(E) \ln[Pr(E)]$$

1.4.3 *The Isothermal Isobaric System; the T, P, N Ensemble*

We now proceed to the next change of a thermodynamic variable. The steps are similar to the ones in Section 1.4.2. Therefore, we shall be very brief.

We start with an ensemble of systems as in Fig. 1.3. Each system has a fixed *volume* V. Remember, in the previous section we started with systems having fixed value of E, and we "opened" the boundaries of the systems so that heat could flow from one system to another. Similarly, we now "open" the boundary of the system so that "volume can flow" from one system to another. What that means is that we replace the solid, rigid boundaries by some flexible boundaries, as shown in Fig. 1.4. The total volume of the ensemble is again fixed, but the volume of each system in the ensemble can fluctuate.

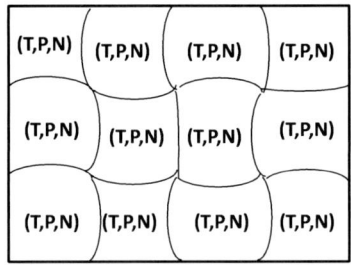

Fig. 1.4 A (T, P, N) ensemble is obtained by replacing the rigid walls by movable boundaries between the systems (see Fig. 1.5).

Therefore, we can talk about the distribution of volumes. For simplicity assume that the volume of the system can attain only discrete values say $V_0, V_1, V_2, \ldots$. We can ask what the probability of finding a single system in the ensemble with a specific volume V is.

The answer to this question is

$$Pr(V) = \frac{Q(T, V, N) \exp[-\beta PV]}{\Delta} \tag{1.20}$$

where Δ is a new normalization constant, which is determined by the condition

$$\sum_V Pr(V) = 1 \tag{1.21}$$

Hence, from Eqs (1.20) and (1.21) we obtain

$$\Delta(T, P, N) = \sum_V Q(T, V, N) \exp[-\beta PV] \tag{1.22}$$

Here, the sum is over all the discrete values of the volume. If the volume changes continuously the sum in Eq. (1.22) is replaced by an integral, and the integral is the Laplace transform of $Q(T, V, N)$ with respect to the variable V. Again, we note that V is used both as the exact volume of each system in the T, V, N ensemble, and as a running variable in Eq. (1.22).

Note carefully how we obtained the change of variables in Eq. (1.22). We started with $Q(T, V, N)$, multiplied by $\exp[-\beta PV]$ and summed over all possible values of V. Therefore, the sum is no longer a function of the parameter V, but of the newly added parameter P, where P is the thermodynamic

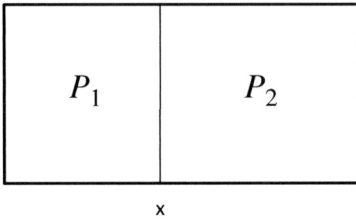

Fig. 1.5 Two systems that are connected by a movable piston. At equilibrium, the pressures on the two sides of the piston are equal.

pressure of the system. The physical reasoning is similar to the one given in the previous subsection. When two systems can exchange volumes, say having a movable piston between them, as shown in Fig. 1.5. The system will reach an equilibrium state at which the *pressure P* on the two sides of the piston will be equal. The same is true for many systems, which can exchange volume. In such an ensemble, volume can "flow" between the systems, but at equilibrium the pressure is constant throughout the entire ensemble.

The fundamental relationship between the new partition function $\Delta(T, V, N)$ and thermodynamics is

$$G(T, P, N) = -k_B T \ln \Delta(T, P, N) \qquad (1.23)$$

where G is the Gibbs energy, expressed as a function of the variables T, P, N. Again we note that this relationship may be established without any additional postulates. As an exercise you can repeat the steps of Eq. (1.15) to obtain this relationship.

Once we have the partition function $\Delta(T, P, N)$ we can use the fundamental relationship as in Eq. (1.23) to derive all the thermodynamic quantities of the system. We use the total differential of G, viewed as a function of T, P, N

$$dG = -SdT + VdP + \mu dN \qquad (1.24)$$

to obtain all the other thermodynamic quantities.

For instance

$$S = -\frac{\partial G}{\partial T} = k_B \ln\Delta + k_B T \frac{\partial \ln\Delta}{\partial T} \qquad (1.25)$$

$$H = G + TS = k_B T^2 \frac{\partial \ln\Delta}{\partial T} \qquad (1.26)$$

Exercise: Show that the entropy of a system in the T, P, N ensemble may be written as

$$S = -k_B \sum_V Pr(V) \ln[Pr(V)]$$

where $Pr(V)$ is given in (1.20).

1.4.4 *The Open System; the T, V, μ Ensemble*

In the previous section we transformed the variable $T, V,$ N to T, P, N (i.e. from V to P). Here we start again with the T, V, N ensemble, and open the systems to a flow of particles, as shown in Fig. 1.6. Clearly, once we replace the

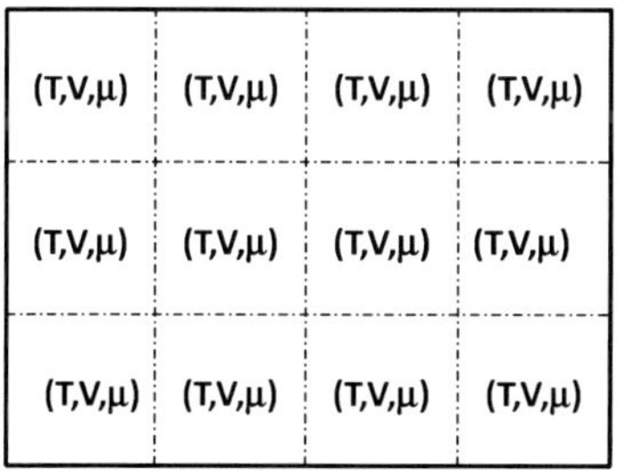

Fig. 1.6 A (T, V, μ) ensemble. The boundaries between the systems are permeable to flow of matter between the systems.

non-permeable boundaries by permeable ones, particles can flow from one system to another. The total number of particles in the entire ensemble is still fixed, but in each system the number of particles will fluctuate (see Section 1.5). In this ensemble we may ask for the probability of finding a system with exactly N particles. The answer to this question is

$$Pr(N) = \frac{Q(T, V, N) \exp[\beta \mu N]}{\Xi} \qquad (1.27)$$

where μ is the chemical potential (assuming there is only one-component in the system), and Ξ is the normalization constant, which is determined by the condition

$$\sum_{N=0}^{\infty} Pr(N) = 1 \qquad (1.28)$$

Hence we obtain

$$\Xi(T, V, \mu) = \sum_{N} Q(T, V, N) \exp[\beta \mu N] \qquad (1.29)$$

Note carefully that in Eq. (1.29) we have multiplied $Q(T, V, N)$ by $\exp[\beta \mu N]$, and hence have introduced a new variable μ. Then we have summed over all possible N, hence the resulting sum is no longer a function of the variable N, but a function of μ, which together with T, V make a new set of variables T, V, μ. As we noted before, the symbol N is used both as the exact number of particles in each system in the T, V, N ensemble, and as a running index in Eq. (1.29).

The quantity $\Xi(T, V, \mu)$ is referred to as the *Grand partition function* and the corresponding ensemble is referred to as the T, V, μ (or the Grand) ensemble.

The fundamental relationship between the Grand partition function and thermodynamics is

$$P(T, V, \mu) = \frac{k_B T}{V} \ln \Xi(T, V, \mu) \qquad (1.30)$$

where P is the pressure of the system characterized by the variables T, V, μ.

How did we get this relationship? The answer is not simple, but let me present a quick argument similar to the ones provided in previous subsections.

We start from Eq. (1.29) and follow the steps given below.

$$
\begin{aligned}
\Xi(T, V, \mu) &= \sum_N Q(T, V, N) \exp[\beta \mu N] \\
&= \sum_N \exp\{-\beta[A(T, V, N) - \mu N]\} \\
&= \sum_N \exp[\beta P(T, V, \mu) V] \\
&\approx \exp[\beta P(T, V, \mu) V] \qquad (1.31)
\end{aligned}
$$

where we have used the thermodynamic relationship

$$G = \mu N = A + PV \qquad (1.32)$$

In the last step of Eq. (1.31) we replace a sum of a large number of positive terms by the maximal term in the sum. This is not a trivial approximation, but for macroscopic systems one can show that this is a good approximation.[3]

As before, once we have constructed the grand partition function $\Xi(T, V, \mu)$, we can proceed with thermodynamics to obtain all other thermodynamic quantities of interest. The appropriate total differential for this case is

$$d(PV) = SdT + PdV + Nd\mu \qquad (1.33)$$

From Eq. (1.33), we can calculate the entropy, the pressure and the average number of particles (Note Eq. 1.30).

$$S = \frac{\partial(PV)}{\partial T} = k_B \ln \Xi + k_B T \frac{\partial \ln \Xi}{\partial T} \qquad (1.34)$$

$$P = \frac{\partial(PV)}{\partial V} = k_B T \frac{\partial \ln \Xi}{\partial V} = k_B T \frac{\ln \Xi}{V} \qquad (1.35)$$

$$N = \frac{\partial(PV)}{\partial \mu} = k_B T \frac{\partial \ln \Xi}{\partial \mu} \qquad (1.36)$$

Before concluding this section we should mention that until now we have assumed that we have a one-component system. In general, the systems can contain any number of components, with composition $N_1, \ldots, N_c$, where N_i is the number of molecules of the type i. If these c-components are independent (i.e. there are no chemical reactions between some of the components), then we must describe each system by the variable T, V, $N_1, \ldots, N_c$. As a shorthand notation we define the vector $\mathbf{N} = (N_1, \ldots, N_c)$ and describe the system by the variables T, V, $\mathbf{N}$. Similarly, in the grand ensemble, the system is described by the variables T, V, $\mu_1, \ldots, \mu_c$, where μ_i is the chemical potential of the ith component.

We define the vector $\boldsymbol{\mu} = (\mu_1, \ldots, \mu_c)$, and the scalar product

$$N \cdot \boldsymbol{\mu} = \sum_{i=1}^{c} N_i \mu_i \tag{1.37}$$

In a multi-component system described by T, V, N, we can "open" the system to one or more of the components. For instance, in a two-component system of A and B, we can open the system either to A or to B. The corresponding grand partition functions are:

$$\Xi(T, V, \mu_A, N_B) = \sum_{N_A} Q(T, V, N_A, N_B) \exp[\beta \mu_A N_A] \tag{1.38}$$

$$\Xi(T, V, N_A, \mu_B) = \sum_{N_B} Q(T, V, N_A, N_B) \exp[\beta \mu_B N_B] \tag{1.39}$$

In the first case we replaced the variable N_A by μ_A, in the second, we replaced N_B by μ_B. These kinds of Grand partition functions are useful in the study of osmotic systems, where one-component can flow through the boundary of the system, but the other component cannot.

Of course, we can always open the system to all the components, the resulting Grand partition function is

$$\Xi(T, V, \boldsymbol{\mu}) = \sum_{N_A} \sum_{N_2} \cdots \sum_{N_c} Q(T, V, N) \exp[\beta \boldsymbol{\mu} \cdot N] \tag{1.40}$$

Exercise: Show that the entropy of a system in the T, V, μ ensemble may be written as

$$S = -k_B \sum Pr(N) \ln Pr(N)$$

where $Pr(N)$ is defined in (1.27).

1.4.5 *The Generalized Partition Function*

In the previous subsections we started with a system characterized by E, V, N, and we successively "opened" the system to the flow of heat, of volume, and of the number of particles. What happens when we "open" the system to all of these variables? Clearly, such an ensemble will be characterized by the variables T, P, $\boldsymbol{\mu}$, i.e. T replaces E, P replaces V and μ_i replaces N_i. Can we find a new partition function, say $f(T, P, \boldsymbol{\mu})$ which is related to some thermodynamic function, as we have done in the previous cases? The answer is no. The reason is the following: Suppose that we could find such a thermodynamic quantity, call it X which is related to the partition function $f(T, P, \boldsymbol{\mu})$, i.e. we could have some relationship of the form

$$X(T, P, \boldsymbol{\mu}) = -k_B T \ln f(T, P, \boldsymbol{\mu}) \qquad (1.41)$$

Let us write the total difference of X:

$$dX = \frac{\partial X}{\partial T} dT + \frac{\partial X}{\partial P} dP + \sum_i \frac{\partial X}{\partial \mu_i} d\mu_i \qquad (1.42)$$

We know from thermodynamics that the variables T, P, $\boldsymbol{\mu}$ cannot be *independent*. The *Gibbs Duhem* equation states that

$$SdT - VdP + \sum_i N_i d\mu_i = 0 \qquad (1.43)$$

This means that there exists an equation of state which connects these variables. Therefore, any thermodynamic function of these variables $X(T, P, \boldsymbol{\mu})$ must be identically zero, i.e. $X(T, P, \boldsymbol{\mu}) = 0$. It follows that we can express any one of the variables T, P, $\boldsymbol{\mu}$ in terms of the others.

Thus, we find that there exists no such partition function $f(T, P, \boldsymbol{\mu})$ which is related to a thermodynamic function X by an equation of the form, as in Eq. (1.41). Nevertheless, the (non-existent) function $f(T, P, \boldsymbol{\mu})$ is referred to as the Generalized partition function, not because it is useful, it does not even exist, but because the procedure of transforming all the variables E, V, N into T, P, $\boldsymbol{\mu}$ produces a useful equation of state. We shall learn more about this peculiar "partition function" in Chapter 9.

1.4.6 *Summary*

As stated in Section 1.1, the aim of statistical thermodynamics is to calculate thermodynamic quantities of macroscopic systems composed of a very large number of microscopic particles. We have started with an isolated system, characterized by a fixed energy E, volume V and number of particles N (or *N*, if there are *c*-components). Such a system is of no interest to us. Any measurement or any observation would necessarily violate the condition of *isolation* of the system. Nevertheless, such a system, or the corresponding ensemble, is the *cornerstone* of statistical thermodynamics. The reason is perhaps paradoxical; we do not know anything about the distribution of the states of such a system. Therefore, the best guess we can make is that all the states of such a system are equally probable. This was one of our postulates. For the moment we will accept this postulate. In addition we have to accept the definition of Boltzmann entropy in Eq. (1.7). We shall see a more profound meaning of the entropy in Chapter 3.

Once we accept the relationship in Eq. (1.7) we can successively transform some or all of the variables E, V, N to obtain a new set of independent variables, which suit our needs. For instance, if we study a chemical reaction in a closed box in a thermostat, it is convenient to choose our independent variables as T, V, N. On the other hand, if our system is still in a thermostat (constant T), but also maintained at constant pressure (say 1 atmosphere), it would be convenient to choose the independent variables T, P, N and so forth. For each set of variables we can construct the corresponding partition function, and the corresponding relationship to a thermodynamic quantity. To do so we do not need to introduce additional postulates; at each step we derive the new relationship from the older one, and ultimately the entire edifice of statistical thermodynamics is built upon the fundamental pillar, which we have referred to as the Boltzmann equation for the entropy. There are of course many mathematical details which are cumbersome and far from trivial, but I want to spare you the trouble of going through the lengthy and elaborate details. Instead you should trust that these details were worked out by outstanding scientists, and also trust that the results obtained from these relationships agree with experimental findings, and therefore these results are trustful.

1.5 Averages and Fluctuations

In this section we present a few average quantities in various ensembles. Consider first the passage from the E, V, N to the T, V, N ensemble. We started with a system having an exact value of the energy E, and "opened" it to allow the flow of heat between the systems.

From Eq. (1.17) we obtain

$$S = -\frac{\partial A}{\partial T} = k_B \ln Q + k_B T \frac{\partial \ln Q}{\partial T} \qquad (1.44)$$

Together with the equation

$$A = \langle E \rangle - TS \qquad (1.45)$$

we have

$$\langle E \rangle = k_B T^2 \frac{\partial \ln Q}{\partial T} = k_B T^2 \frac{\sum_i \exp[-\beta E_i]}{kT^2 Q}$$

$$= \sum_i E_i Pr(E_i) = \sum_E EPr(E) \qquad (1.46)$$

$\langle E \rangle$ is the average energy of a system in a T, V, N ensemble. Note again that the symbol E is used both as the exact energy of a single system in the E, V, N ensemble, and as a running variable in Eq. (1.46).

Note carefully that $Pr(E_i)$ is the probability of finding a system in a state i and the sum over i is the sum over all *states* of the system. If the energy level E_i has degeneracy Ω_i, we can rewrite the average also as a sum over all energy levels. In each case we get the average energy of the system. Since the total energy of the ensemble is fixed, we must have

$$Total\ energy = ME = M\langle E\rangle \qquad (1.47)$$

Equivalently

$$E = \langle E\rangle \qquad (1.48)$$

This result is what we expect to get. We start with an ensemble of isolated systems, each having an exact energy E. We "opened" the systems to heat exchange among all the systems, but the total energy of the ensemble was not changed. The energy of each system can change, but its *average* energy is equal to the exact energy we had in the original ensemble.

Exercise: Repeat the same procedure as above for the process of opening the T, V, N ensemble to exchange of volume. Show that the average volume of a system in the T, P, N ensemble is

$$\langle V\rangle = \sum_V VPr(V) \qquad (1.49)$$

where V is the exact volume of each system in the T, V, N ensemble, and $\langle V\rangle$ is the average volume in the T, P, N ensemble. In the sum Eq. (1.49) V is a running variable.

Exercise: Repeat the same procedure for the process of opening the T, V, N ensemble to flow of particles, and show that

$$\langle N\rangle = \sum_N NPr(N) \qquad (1.50)$$

Note again that N is the *exact* number of particles of a system in the T, V, N ensemble, and $\langle N\rangle$ is the average number of particles in the T, V, μ ensemble. In the sum Eq. (1.50) N is a running variable.

Other important average quantities are the fluctuations. Clearly, for an isolated system the energy E is fixed. However, when we "open" the systems to heat flow, we are interested not only in the *average* energy of each system, but also in a measure of the average deviation from the average energy of the system.

The extent of fluctuations is measured by the standard deviation σ_E defined by

$$\sigma_E^2 = \langle (E - \langle E \rangle)^2 \rangle \tag{1.51}$$

We can use the probability distribution $Pr(E)$ to calculate this average

$$\sigma_E^2 = \sum_E (E - \langle E \rangle)^2 Pr(E)$$

$$= \sum_E E^2 Pr(E) - 2E\langle E \rangle Pr(E) + \langle E \rangle^2 Pr(E)$$

$$= \langle E^2 \rangle - 2\langle E \rangle \langle E \rangle + \langle E \rangle^2 = \langle E^2 \rangle - \langle E \rangle^2 \tag{1.52}$$

On the other hand if we differentiate the average energy in Eq. (1.46) with respect to the temperature, we obtain the heat capacity

$$C_V = \frac{\partial \langle E \rangle}{\partial T}$$

$$= \frac{Q \sum_E \frac{E^2}{k_B T^2} \exp[-\beta E] - \sum E \exp[-\beta E] \sum \frac{E}{k_B T^2} \exp[-\beta E]}{Q^2}$$

$$= \frac{1}{k_B T^2} (\langle E^2 \rangle - \langle E \rangle^2) \tag{1.53}$$

Hence, we obtain the important result

$$\sigma_E^2 = k_B T^2 C_V \tag{1.54}$$

Thus, the constant volume heat capacity is related to the average fluctuation in the energy of the systems in the T, V, N ensemble.

Note that by definition, σ_E^2 is a positive quantity. It follows that C_V is always a positive quantity.

Furthermore, if we take the relative deviations

$$\frac{\sigma_E}{\langle E \rangle} = \frac{(k_B T^2 C_V)^{1/2}}{\langle E \rangle} = \frac{O(\sqrt{N})}{O(N)} = O\left(\frac{1}{\sqrt{N}}\right) \tag{1.55}$$

Thus, the relative deviation changes with N as $1/\sqrt{N}$.

Exercise: In transforming from the variables T, V, N to T, P, N, the exact volume V is replaced by the average volume $\langle V \rangle$. Calculate the standard deviation σ_V and show that in the T, P, N ensemble

$$\sigma_V^2 \equiv \langle V^2 \rangle - \langle V \rangle^2 = -k_B T \left(\frac{\partial \langle V \rangle}{\partial P}\right)_{T,N} \tag{1.56}$$

$$\kappa_T \equiv \frac{-1}{\langle V \rangle} \left(\frac{\partial \langle V \rangle}{\partial P}\right)_{T,N} = \frac{\langle V^2 \rangle - \langle V \rangle^2}{k_B T \langle V \rangle} \tag{1.57}$$

Similarly

$$C_p \equiv \left(\frac{\partial \langle H \rangle}{\partial T}\right)_{P,N} = \frac{\langle H^2 \rangle - \langle H \rangle^2}{k_B T^2} \tag{1.58}$$

$$\alpha \equiv \frac{1}{\langle V \rangle} \left(\frac{\partial \langle V \rangle}{\partial T}\right)_{P,N} = \frac{\langle VH \rangle - \langle V \rangle \langle H \rangle}{k_B T^2 \langle V \rangle} \tag{1.59}$$

where κ_T is the *isothermal compressibility*, C_p is the *constant pressure heat capacity*, and α is the *thermal expansion coefficient*.

Exercise: Calculate the fluctuations in the number of particles in the grand ensemble.

Follow almost the same procedure as we did in calculating the fluctuations in energy, to derive an expression for the fluctuations in the number of particles in the T, V, μ ensemble

$$\sigma_N^2 \equiv \langle (N - \langle N \rangle)^2 \rangle$$

$$= \langle N^2 \rangle - \langle N \rangle^2 = k_B T V \left(\frac{\partial \rho}{\partial \mu} \right)_T \qquad (1.60)$$

where $\rho = \langle N \rangle / V$ is the average density of particles in a system in the T, V, μ ensemble.

Show also that the isothermal compressibility may be written as[4]

$$\kappa_T = \frac{\sigma_N^2}{k_B T \rho^2 V} \qquad (1.61)$$

1.6 Classical Statistical Thermodynamics

In this chapter, we introduced the second postulate of statistical thermodynamics, and the various ensembles in statistical thermodynamics, using the language of quantum mechanics. All we had to know is that a system can be described by its quantum mechanical state Ψ which is a solution of the Schrödinger time-independent equation. We do not need to solve the Schrödinger equation for any macroscopic system. All we have to assume is that there exist energy levels (which are the eigenvalues of the

Schrödinger equation), and that each energy level has a degeneracy Ω_E.

In most applications of statistical thermodynamics for problems in Chemistry and Biochemistry, it is more convenient to describe a system *classically*, i.e. the *state* of a system is described by giving the locations and the momenta of all the atoms, or molecules of the system. With this description we introduce the classical canonical partition function (PF) $Q(T, V, N)$ by analogy. The energy levels are replaced by the classical Hamiltonian function, i.e.

$$E \leftrightarrow H(p^N, R^N) = \sum_{i=1}^{N} \frac{p_i^2}{2m} + U_N(R^N) \qquad (1.62)$$

Thus, for each set of parameters $p_1, p_2, \ldots, p_N$ and $R_1, R_2, \ldots, R_N$ there corresponds an energy "level" which consists of the kinetic energy of all the particles and the potential energy of interaction between all the particles.

The "classical" analog of the canonical PF is thus

$$Q(T, V, N) \approx \int \cdots \int dp^N dR^N \exp[-\beta H(p^N, R^N)]$$

$$(1.63)$$

Here, the sum over all quantum mechanical *states* of the system is replaced by integrations over all positions and momenta of the particles. It turned out that this PF *does not* lead to results that conform with the experimental results. To achieve such an agreement we must add two factors, the origin of both is in quantum mechanics.

In Chapter 2, we shall see the origin and the meaning of these two terms. Here, we just quote the final result for the "corrected" classical partition function in Eq. (1.63).

$$Q(T, V, N) = \frac{q^N}{(8\pi^2)^N \Lambda^{3N} N!} \int dX^N \exp[-\beta U_N(X^N)]$$

(1.64)

where q is the internal partition function of a single molecule, Λ^{3N} is a result of integration over all possible momenta of the particles

$$\Lambda^3 = \left(\frac{h}{\sqrt{2\pi m k_B T}} \right)^{3/2}$$

(1.65)

In Eq. (1.65) h is the Planck constant and k_B is the Boltzmann constant. $X^N = X_1, \ldots, X_N$ is the *configuration* of all the N particles, with X_i being the vector comprising the location and orientation of the ith particle. In Chapter 2, we shall see how the Planck constant h and the $N!$ enter into the entropy of an ideal gas.

2

Shannon's Measure of Information

2.1 Introduction

This chapter introduces a concept, which is not ordinarily used in statistical thermodynamics or in thermodynamics. Nevertheless, I believe that a student of any of the sciences should be familiar with this concept. In the context of this book it is used for one reason only; to provide a meaningful interpretation of entropy.

In Section 1.4 we have defined the Boltzmann entropy by the relationship

$$S = k_B \ln \Omega \qquad (2.1)$$

where k_B is the Boltzmann constant and Ω is the number of quantum mechanical states of an isolated system characterized by E, V, N. In Section 1.4 we saw that in any ensemble the entropy may be expressed in terms of the fundamental probability distribution in that ensemble, i.e.

$$S = -k_B \sum_i p_i \ln p_i \qquad (2.2)$$

where $(p_1, p_2, \ldots)$ is the probability distribution of energies in the T, V, N ensemble of the volumes in the T, P, N ensemble or of the number of particles in the T, V, μ ensemble.

We now wish to generalize the concept of entropy as expressed in Eq. 2.2 to *any* probability distribution.

To do that we first eliminate the constant k_B, and also we shall use the base 2 for the logarithm. Thus, following Shannon,[1] we *define* the quantity

$$H = -\sum_{i=1}^{n} p_i \log_2 p_i \qquad (2.3)$$

for any distribution $(p_1, \ldots, p_n)$. Here, for simplicity we define H for a finite number of possible outcomes, but the definition may be extended to an infinite number of outcomes or even to a continuous range of outcomes. We shall use the definition of H for any set of numbers which we shall refer to as a *probability distribution*, i.e. the numbers p_i satisfying the conditions

$$0 \le p_i \le 1, \quad \sum_{i=1}^{n} p_i = 1 \qquad (2.4)$$

Imagine an experiment, or a game with n possible outcomes, the probability of occurrence of the ith outcome being p_i. For instance, throwing a dice has six possible outcomes $(1, 2, 3, 4, 5, 6)$. In general, the probability distribution of a dice could be any set of six numbers fulfilling the conditions (2.4). A fair (or an "honest," or a symmetric) dice is defined as one having a uniform probability distribution, i.e. $(\frac{1}{6}, \frac{1}{6}, \frac{1}{6}, \frac{1}{6}, \frac{1}{6}, \frac{1}{6})$.

Suppose that we play with a dice. You make a guess on what the outcome will be when I throw the dice. If the outcome is the same as the one you had guessed, you win, otherwise you lose.

Now suppose that I give you the following *information*: I tell you that the distribution of the outcomes is, say

Outcomes:	$(1, 2, 3, 4, 5, 6)$	
Probabilities:	$(0.1, 0.2, 0.3, 0.2, 0.1, 0.1)$	(2.5)

Does that help you in guessing the right outcome? You correctly feel that to some extent it does help. Why? Because if you guess that the outcome is 3, there is a better chance that you will win. Better in what sense? Clearly, you can guess the outcome 3 and the actual outcome might be 1, 2, 4, 5 or 6, and you will lose. To clarify the question of how much the information that I provided helped, consider the following two extreme cases: In one case, call it (a), I give you the distribution

$$p_a = (1, 0, 0, 0, 0, 0) \qquad (2.6)$$

And in the second case, I give you the distribution (b)

$$p_b = \left(\frac{1}{6}, \frac{1}{6}, \frac{1}{6}, \frac{1}{6}, \frac{1}{6}, \frac{1}{6} \right) \qquad (2.7)$$

Clearly, you would not hesitate to say that in case (a) the information I gave you was the "best," whereas in the second case (b) it is the "worst." In the first case the information I gave you about the distribution is equivalent to the information on which outcome you should bet in order to win. In the second example the information I gave you on the distribution is *useless*; it cannot improve your chances of winning. In example 2.5 the amount of information I gave you in the distribution is somewhere between these two extremes: It is not as good as in (a), but it is not as useless as in (b).

The assertion that the "quality" of the information contained in the distribution (2.5), is somewhere between (a) and (b) is easily understood. But it would not be easy to compare the distribution (2.5) with the following one

$$(0.2, 0.3, 0.4, 0.05, 0.05, 0) \tag{2.8}$$

Our aim is to find a quantitative *measure* of the amount of information contained in, or associated with a given distribution.

2.2 The 20-Questions (20-Q) Game

Consider the following simple game. You are shown an array of say, 32 objects or persons, as in Fig. 2.1. I am thinking of one object and you have to find out which object I am thinking of. This is the familiar game, known as the 20-question game. In fact this is already a *reduced* 20-Q game. Normally, in the 20-Q game, one player chooses either an object or a person, and the second player has to find out which object or person was chosen by the first player by asking binary questions (i.e. questions that can be answered by YES or NO). The game in Fig. 2.1 is a *reduced* game, in the sense that I am allowed to choose only one object from a given set of objects, or a given set of persons. We shall soon further

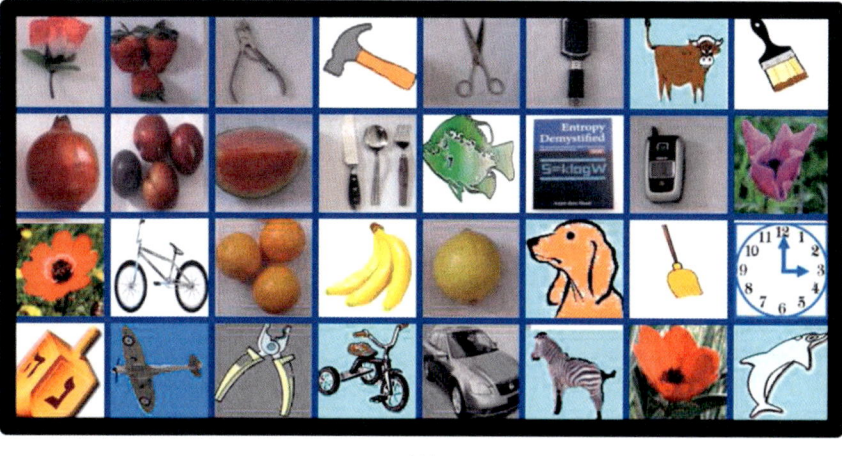

(A)

(B)

Fig. 2.1 An array of 32 objects [(A) and persons (B)] used in the 20-Q game.

reduce the game to make it more precise. However, before we do that it is clear that the problem we face in this game is to acquire some information i.e. "which is the object I chose." The main objective of this game is not just to *find* the missing information, but to find it in the most "*efficient*" way.

To quantify the term "efficient" in this context, suppose that I chose an object from the 32 objects in Fig. 2.1A, and you have to find out the object by asking binary questions. Once you find the object I chose, you get a prize, say 10 USD. Suppose also that you have to pay, say one dollar for each answer you get. Assuming that you want to maximize your gain in playing this game, you should do your best to ask the minimum number of questions. This is equivalent to maximizing your earnings (i.e. the prize of 10 USD minus the total number of dollars you have paid in the process of questioning). Thus, in the context of this game *efficiency* means earnings, and *maximal* efficiency means maximal earning, or minimal number of questions.

Intuitively, you feel that there are different *ways* of asking questions. For instance, you can ask; is it a banana? is it a book? is it a dog, etc. You also feel that this *strategy* of asking questions would not be efficient. We shall later see that if you adopt this strategy, and play the game many times you will, on the average, *lose* money.

Another way of asking questions is to ask: Is it a tool, or is it an animal, or a red object? Adopting this strategy is somewhat more efficient, but still short of being the *most* efficient strategy.

As we shall soon see there exists a strategy, in this particular game, such that if you adopt it, you will obtain the required information in exactly five questions. This means that if you are smart enough to find out the smartest strategy, you are guaranteed to gain in every game you play (i.e. you will pay 5 USD for the answers, and you will earn the 10 USD prize upon finding the object). Most amazingly, information theory provides a precise mathematical method to finding the *smartest* strategy. We shall see this in the following sections.

2.3 Definition of Shannon's Measure of Information for a Uniformly Distributed 20-Questions Game

In the previous section we discussed a simple version of the well-known 20-Q game. It is a simple version in the sense that we have specified exactly the set of objects from which we choose one object. In the traditional 20-Q game one chooses a person or an object without specifying the complete set of persons or objects. In playing the traditional game we do not make any statements regarding the relative probabilities of choosing one person over another one, or one object over another.

We shall now *formalize* the 20-Q game in such a way that we can define a precise *measure* on the game. This measure will be applied in the next section to a more general game.

Consider the following game: There are N *identical* boxes. I hide a coin in one of the boxes and you have to find out the box in which the coin is hidden. I also inform you that I have chosen the box in which I placed the coin at *random*. Qualitatively, this means that I have no *preferred* box and the likelihood of choosing any specific box is simply $1/N$.

In this version of the game the task is to find out in which box the coin is hidden. If the boxes were labeled, say by numbers: $1, 2, 3, \ldots, N$, and if I placed the coin in the 4th box, then I know the *information* about the location of the coin, but you do not have that *information*. The information

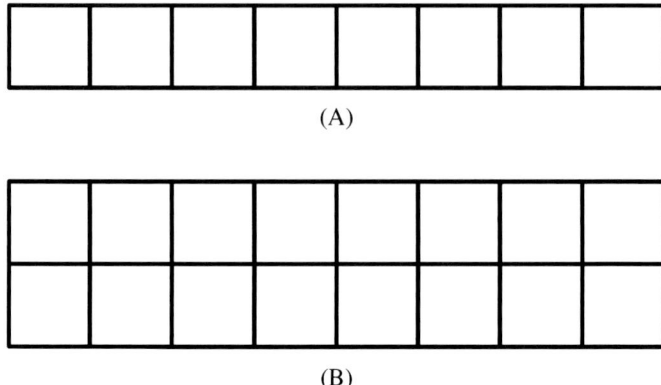

Fig. 2.2 Two games with (A) eight boxes and (B) sixteen boxes.

is: "the coin is hidden in the 4th box." The task of the game is to find this *information*. The task of *Information Theory* is to find a *measure* of the *size of this task*.

Before we introduce the measure of the task or the measure of the game, consider the two games in Fig. 2.2. In the first there are eight boxes, and I hid the coin in box ($1 \leq k \leq 8$), which was chosen at *random*. In the second game there are 16 boxes, and I placed the coin in one randomly chosen box, say $k'(1 \leq k' \leq 16)$. I know the information on the location of the coin in each game, but you lack that information. In this particular game, Information Theory is not concerned with the *content*, or with the *meaning* of this information, but rather with the question of how easy or how difficult it is to find out this information by asking binary questions. It is intuitively clear that the first game is easier than the second. Easier, in the sense that the minimal number of binary questions one needs to ask to acquire the missing information is smaller than in the second game.

The last sentence needs some clarification. We noted in the previous section that there are many different *strategies* of asking questions. Let us consider two extreme cases. Suppose we adopt the following strategy, i.e. we ask: Is the coin in box 1? If the answer is NO, we ask: Is the coin in box 2, and so on until we get a YES answer. For reasons that will be clear later we shall refer to this strategy as the "dumbest" strategy.

Clearly adopting this strategy for the two games in Fig. 2.2 we feel that we shall need more questions *on average* in the second game than in the first one.

I said "on average," because you might play any one of the two games, and win the prize after the first question. However, the probability of winning the first game on the first question is 1/8, and 1/16 in the second game. Therefore, if we play these two games many times by adopting the "dumbest" strategy, we shall need more questions *on average* to find the coin in the second game than in the first game.

The second strategy, which we shall refer to as the "smartest," is to divide the total number of boxes into two equal parts and ask: Is the coin in the left hand side group of boxes? If the answer is YES, we divide them again into two equal groups, and ask the same question. If the answer is NO, we ask the same question but on the group of boxes on the right hand side. The two strategies are illustrated in Figs. 2.3A and 2.3B.

Clearly, if we adopt this strategy we shall find the required information in exactly three questions in the first game, Fig. 2.3A, and in exactly four questions in the second game, Fig. 2.3B. We find again that the second game is more

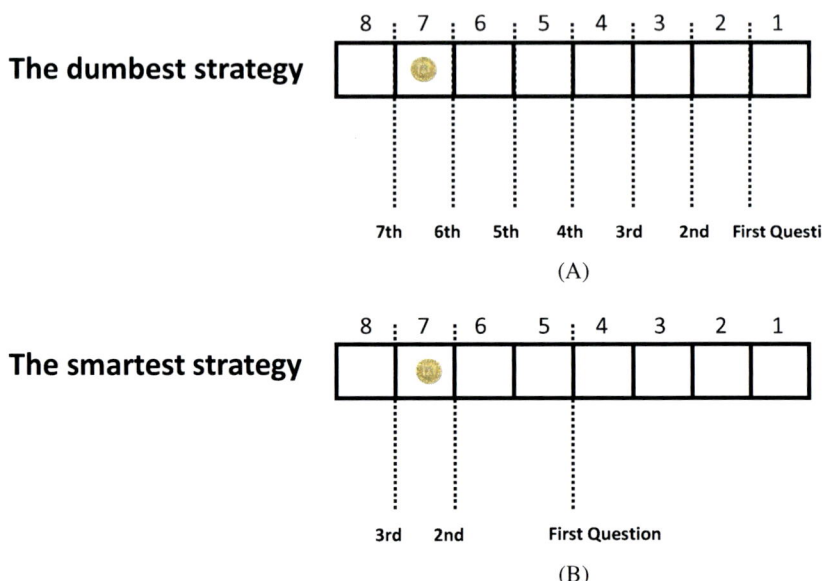

Fig. 2.3 Two strategies of asking questions; (A) the "dumbest" strategy and (B) the "smartest" strategy.

"difficult" than the first, in the sense that we need more questions to ask in the second, compared with the first.

Two conclusions may be derived from the above examples:

First, for any specific game, either (A) or (B) in Fig. 2.3, the "smartest" strategy is more "efficient" than the "dumbest" one. It is more efficient in the sense that "on average" we shall need to ask fewer questions to obtain the required information. We shall also refer to these kinds of questions as "smart questions."

The second conclusion is that the game with the larger number of boxes requires more questions in order to find the coin, compared to the game with the fewer number of boxes.

These two conclusions were derived for the specific examples in Fig. 2.3. It turns out that these conclusions have more general validity. Here, we provide some qualitative plausible arguments. Suppose we have a game with N equally probable boxes (that means that we have selected the box in which to place the coin at random, given that each box has the same "chances" or the same probability). Suppose that N is of the form

$$N = 2^n \qquad\qquad (2.9)$$

where n is an integer. In this case the two strategies of asking questions are:

Dumbest:

1. Is the coin in box number 1?
2. Is the coin in box number 2?

    and so on. . .

Smartest: At each step we divide the total number of boxes into exactly two halves (this is possible, because we have chosen N to be of the form 2^n as in 2.9), and ask: Is the coin in the right hand side (rhs) or the left hand side (lhs) of the boxes?

    Intuitively, it is clear that by using the smartest strategy we shall find the coin in exactly n steps. The reason is simple: at each step we divide into two halves, hence the sizes of the two groups of boxes at each step are:

Initial number: $N = 2^n$
First step: $(N/2 = 2^{n-1}, N/2 = 2^{n-1})$
Second step: $(N/4 = 2^{n-2}, N/4 = 2^{n-2})$

$$\vdots$$

$(n\text{-}1)$th step: $(1,1)$
nth step: we find the coin

Thus, for $N = 2^n$ we shall find the coin in exactly n steps, which is given by

$$n = \log_2 N = \log_2 2^n \qquad (2.10)$$

On the other hand, using the "dumbest" strategy we divide the entire group of boxes at each step as follows:

Initial number: $N = 2^n$
First step: $(1, N - 1)$
If the first step failed
the second step: $(1, N - 2)$
and so on.
Failing the N-2 steps
we reach the N-1th step $(1,1)$
At the Nth step we get the answer.

It is qualitatively clear that in the "smartest" strategy the number of steps (hence, the number of questions) is of the order of $\log_2 N$. On the other hand, in the "dumbest" strategy the number of steps (hence, the *average* number of questions) is of the order of N, more precisely it varies with N as $N/2$.[2] Figure 2.4 shows the average number of questions required to find the coin as a function of N for the two strategies. In both cases the average number of questions grows with N, but it grows much faster in the "dumbest" strategy compared with the "smartest" strategy. This is why we have labeled the two strategies as "smartest" and "dumbest." We shall see in Section 2.4 than when we choose the "smartest" strategy, *we gain more information* at each step than when we choose the "dumbest" strategy. Gaining more information at each step is the same as obtaining the required information in

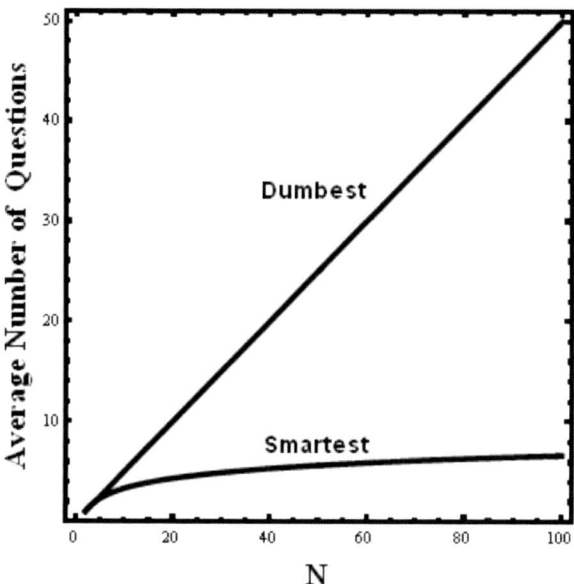

Fig. 2.4 The average number of questions as a function of N for the "dumbest" and the "smartest" strategy.

fewer questions. Of course, there are many strategies that fall in between these two extreme cases. Here, we shall be interested only in the smartest one, which we shall use to *define* the "size" of the game, or the "size" of the required information.

Note that we could have used the number of objects N as a measure of the "size" of the game. The larger the N the harder it is to play the game. We could also have chosen N^2 or N^{10} or $\exp[N]$ to measure the size of the game. However, one choice has a particularly simple interpretation. For any game or any experiment with N equally probable outcomes we define the *Shannon measure* of *information* (SMI) by

$$\text{SMI} = \log_2 N \tag{2.11}$$

As we have seen for the specific case of N of the form 2^n, the SMI is also the number of binary questions we need to ask in the smartest strategy in order to find the box in which the coin is. Or, in the most general case of N possible outcomes of an experiment the SMI is the number of questions we need to ask in order to find out which outcome has occurred.

So far we have discussed only the case of a number of N having the form 2^n (with n being an integer). For any N we can still define the SMI as in Eq. (2.11) but now the quantity SMI is not exactly the number of questions. For any number of boxes N we can always find an integer n, such that N falls between the two consecutive powers of 2, i.e.

$$2^n \leq N \leq 2^{n+1} \tag{2.12}$$

Exercise: Take a few numbers $N = 30, 100, 200$ and convince yourself that you can find an integer n such that Eq. (2.12) is valid. For instance, if N is equal to ten, then

$$8 = 2^3 \leq 10 \leq 2^4 = 16 \tag{2.13}$$

Thus, for any given N boxes we can always add empty boxes until we reach the nearest number of the form 2^n. Doing that will only add one more question. Thus, for very large N we can still interpret the SMI in terms of the number of "smart" questions, neglecting the difference between $n + 1$ and n.

In Section 2.5 we shall generalize the conclusion we have reached in this section to the case of a non-uniform distribution. Before doing so, we still have to clarify one important point: How much information do we gain by asking one question? The next section is devoted to clarifying this point.

2.4 The Case of Two Outcomes

Consider the case $N = 2$. We have only two boxes, in one of them a coin is hidden. You have to find out where the coin is by asking binary questions. Clearly, by asking one question you will *know* where the coin is. This seems to be in accordance with the conclusion of the previous sections, namely that $\log_2 N = \log_2 2 = 1$, i.e. one question needs to be asked. Note however that in this particular case we did not say anything regarding the probabilities of placing the coin in the first, or the second box. Whatever the probabilities are, we must ask one question in order to know where the coin is (we exclude for the moment the case of probability zero for one box, and one for the second).

However, intuitively you feel that if the probability distribution is symmetric [i.e. (½, ½) for the two boxes] you have less *information* than when the probability distribution is asymmetric. To clarify this point, consider a different experiment.

We throw a dart towards a board having an area, say, $A = 1m^2$. The board is divided into two parts of areas A_1 and A_2, respectively, such that $A = A_1 + A_2$ (Fig. 2.5). You are told that the dart has hit the board at some point but not in which one of the two areas A_1 and A_2, and that the dart was thrown at random, meaning that all the points in the board are equivalent. Therefore, the probability of hitting the first area is $p_1 = A_1/A$, and hitting the second area is $p_2 = A_2/A$, with

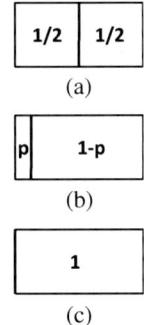

Fig. 2.5 Three possible divisions of a board into two areas (a) symmetric, (b) asymmetric, (c) extreme asymmetric.

$p_1 + p_2 = 1$. Thus, you are given the probability distribution (p_1, p_2), and that the dart hit the board and your task is to find out in which of the two areas the dart is.

Consider the following three cases:

(a) Symmetric distribution: $(½, ½)$
(b) Asymmetric distribution: $(p1 - p)$, $(0 < p < ½)$
(c) Extreme asymmetric distribution: $(0,1)$

These three cases are shown in Fig. 2.5. In the first case you must ask one question in order to know where the dart hit. In the third case you do not have to ask any question; given the distribution $(0, 1)$ is equivalent to having the knowledge that the dart hit the right hand area, i.e. you know where the dart is. In any intermediate case of $0 < p < \frac{1}{2}$, you feel intuitively that you "know" more information than in the first case but less than in the third case. On the other hand, in all cases except the third one, one must ask at least one question in order to find out where the dart is. Thus, we cannot use the "number of questions" to quantify the amount of "information" we have in the general case of $(p, 1 - p)$. To do so we modify the rules of the game, or the task you are facing.

Again, you know that the dart hit the board, and you are given the distribution $(p, 1 - p)$, i.e. you know the areas A_1 and A_2. You choose one area and ask if the dart is in that area. If the answer is YES, you win, and if the answer is NO, you ask: Is the dart in the second area, and the answer you will certainly get is, YES.

The difference between the rules of this game and the previous ones is simple. When you have only two possibilities, you will *know* where the dart is after the first question, whatever the answer is. However, in the modified game, if the answer is NO, we add one more "verification" step, for which the answer is always YES.

Now let us calculate the number of questions in this modified game.

For the symmetric case the probability of obtaining a YES answer on the first step is ½, and of obtaining a NO answer is also ½. Therefore, the average number of questions in the symmetric distribution case is

$$\frac{1}{2} \times 1 + \frac{1}{2} \times 2 = 1\frac{1}{2} \tag{2.14}$$

Thus, on average you need 1.5 questions.

For the extreme asymmetric distribution $(p = 0)$, you will ask exactly *one* question: Is the dart on the right hand side and the answer will be YES. In this case you need only the verification question.

In any intermediate asymmetric case, we have $(p, 1 - p)$, with $0 < p < \frac{1}{2}$. So the first question should be: Is the dart in the rhs? You will obtain a YES answer with probability $(1 - p)$. Therefore, the average number of questions in this case is

$$(1 - p) \times 1 + p \times 2 = 1 + p \qquad (2.15)$$

This result means that for $0 < p < \frac{1}{2}$, the average number of questions is between 1 and 1.5, provided we always start by asking whether the dart is in the area of highest probability (i.e. the one with probability $1 - p$, if $0 < p < \frac{1}{2}$). If we decide to start by asking on the area of the lower probability the average number of questions will be

$$p \times 1 + (1 - p) \times 2 = 2 - p \qquad (2.16)$$

which is larger than (2.15), for $0 < p < \frac{1}{2}$.

Of course, if we are interested in asking the minimal number of questions, we should always choose to ask first about the area of larger probability. With this choice the average number of questions as a function of p is shown in Fig. 2.6. This function conforms with our expectations in the following sense: When p is either zero or one, we *know* where the dart is. Therefore, according to the rules of the previous section we do not need to ask any questions. According to the modified rules of the game in this section we must ask only one question, the question of verification only. On the other hand, when the distribution is symmetric we have no clue about the location of the dart. In this case we must ask more questions: one, where is the dart, and if you got a NO answer, you must ask a second verification question. Thus, on average you will need 1½ questions to obtain the information and verify it.

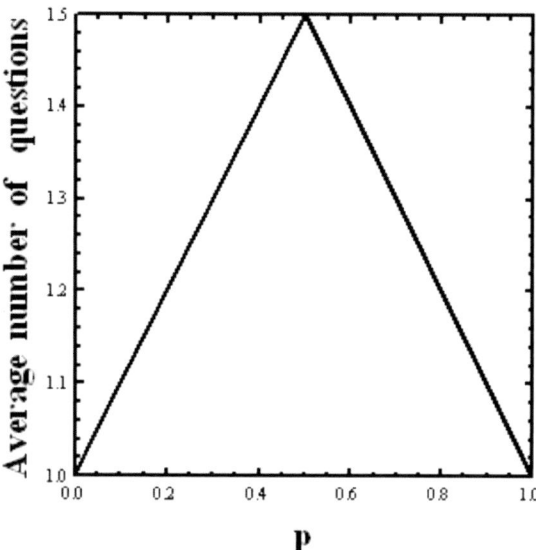

Fig. 2.6 Average number of questions in the modified game of two outcomes.

For any other distribution $(p, 1-p)$ we have an intermediate value between 1 and 1½. This result conforms our intuitive feeling that in this particular game the knowledge of the symmetric distribution does not provide any *information* to help us in locating the dart, therefore we must put in more effort, i.e. ask more questions to find where the dart is. The more asymmetric the distribution, the more *information* we have, hence the less effort is needed to locate the dart. In the extreme distribution case, either $(0, 1)$ or $(1, 0)$, we have all the information we need to locate the dart, therefore we have to ask the minimum number of questions (here, only one; the verification question).

All that we have said above was very qualitative. We could have invented many functions which have a maximum at $p = ½$, and minimum at $p = 0$ and $p = 1$. All these functions conform with our expectation of a measure of the amount of missing information or about the result of the experiment. The more asymmetric the distribution the more certain (or the less uncertain) we are about the outcome of the experiment. Alternatively, we can also say that the more asymmetric the distribution, the smaller is the amount of *missing information* on the outcome of the experiment.

Shannon has defined a more general measure of information, which is valid for any N outcomes and for any distribution of the outcomes. We shall discuss the general definition in the next section. Here, we introduce Shannon's measure for the particular case of two outcomes. This is defined by[3]

$$SMI(p) = -p \log_2 p - (1-p) \log_2 (1-p) \qquad (2.17)$$

The form of this function is shown in Fig. 2.7A. This function has a maximum at $p = \frac{1}{2}$ and a minimum at both

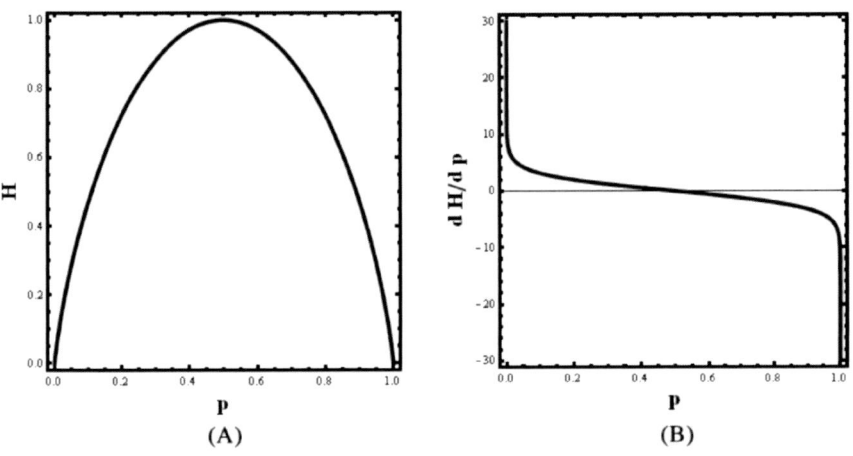

Fig. 2.7 The SMI for the case of two outcomes. (A) and its derivative (B).

$p = 0$ and $p = 1$. We shall further discuss the function SMI (p) for the general case in the next section. Here, we note that if we choose the base 2 for the logarithm, then the maximal value of SMI (p) is

$$SMI\left(p = \frac{1}{2}\right) = -\frac{1}{2}\log_2 \frac{1}{2} - \frac{1}{2}\log_2 \frac{1}{2} = 1 \qquad (2.18)$$

This value is used as the unit of information called *bit* (short for *binary digit*).[4] Other units, less frequently used are: $\ln 2 = \log_e 2 = 0.693$ Nats, or $\log_{10} 2 = 0.301$ Hartleys. As we shall see in Section 2.5 the base 2 is the most convenient choice also for the more general case, i.e. for any number of outcomes and any probability distribution of outcomes.

2.5 Shannon's Measure of Information (SMI) for the General Distribution

In 1948, Shannon published a landmark paper titled, "*A Mathematical Theory of Communication.*" In Section 6 of this paper Shannon writes:

> "*Suppose we have a set of n possible events whose probabilities of occurrence are $p_1, p_2, \ldots, p_n$. These probabilities are known but that is all we know concerning which event will occur. Can we find a measure of how much "choice" is involved in the selection of the event or how uncertain we are of the outcome?*
> *If there is such a measure, say $H(p_1, p_2, \ldots, p_n)$, it is reasonable to require of it the following properties:*
> 1. *H should be continuous in the p_i.*
> 2. *If all the p_i are equal, $p_i = \frac{1}{n}$, then H should be a monotonically increasing function of n. With equally likely events there is more choice, or uncertainty, when there are more possible events.*

3. *If a choice be broken down into two successive choices, the original H should be the weighted sum of the individual values of H.*

Then Shannon proved that:

"The only H satisfying the three assumptions above is of the form:

$$H = -K \sum p_i \log p_i \qquad (2.19)$$

In this book we shall not be interested in the *derivation* of the quantity H, which we will refer to as *Shannon's measure of information* (SMI). In this chapter, we are only interested in the *properties* of SMI, and its *meaning* as a measure of information. The relevance to thermodynamics will be discussed in Chapter 3. Here, we will study the quantity H as defined in Eq. (2.19) without any reference to thermodynamics. Let us quote another paragraph from Shannon's paper.

> *"This theorem, and the assumptions required for its proof, is in no way necessary for the present theory. It is given chiefly to lend a certain plausibility to some of our later definitions. The real justification of these definitions however, will reside in their implications.*
>
> *Quantities of the form $H = -K \sum p_i \log p_i$ (the constant K merely amounts to a choice of a unit of measure) play a central role in information theory as measures of information, choice and uncertainty. The form of H will be recognized as that of entropy as defined in certain formulations of statistical mechanics where p_i is the probability of a system being in cell i of its phase space. H is then, for example, the H in Boltzmann's famous H theorem. We will call $H = -K \sum p_i \log p_i$ the entropy of the set of probabilities $p_1, \ldots, p_n$."*

Note carefully that Shannon describes H as a "measure of information, choice and uncertainty". All these are valid interpretations of the quantity H as defined in Eq. (2.19). Shannon goes on to say that "the *form* of H will be recognized as that of *entropy* as defined in certain formulations of statistical mechanics where p_i is the probability..." The reader is urged to carefully examine the above quotation. You do not need to understand all the details. Note however, a few points that will be crucial for understanding the relevance of the SMI to statistical thermodynamics.

First, note that Shannon formulated his problem in terms of a probability distribution: $p_1, \ldots, p_n$. He sought a measure of how much "choice", or "uncertainty" there is in the outcome, and later he referred to the quantity H as a measure of "information, choice and uncertainty". We will further discuss these meanings of the quantity H in Section 2.6.

Shannon did not seek a measure of the general concept of information, only a *measure* of *information contained in*, or *associated with* a probability distribution.

Second, Shannon proposed three plausible properties of such a measure, *presuming* that such a measure exists. We shall discuss these properties and their plausibility in the following sections of this chapter. Here, we draw the attention of the reader to the "methodology" of seeking and finding a quantity that we are not even sure whether it exists or not.

Finally, note carefully that Shannon was not interested in thermodynamics in general, nor in *entropy* in particular. However, he noted that "the form of H will be recognized as that of entropy as defined in certain formulations of statistical mechanics.." Therefore, he suggested to call H "the entropy of the set of probabilities $p_1, \ldots, p_n$".

Indeed, the *form* of the function H is the same as the *form* of the entropy as used in statistical mechanics. However, the fact that the *form* of H is the same as the form of the entropy in statistical mechanics does not imply that H is entropy. We shall further discuss this point in Chapter 3, after we learn some of the properties of the SMI and of the entropy. For the moment we shall study the SMI without any reference to entropy. However, the reader should be aware of the fact that in many applications of SMI, the concept of entropy has also been involved.

Shannon's measure of information is very general. It is defined on *any distribution*; the outcomes of throwing a dice or the frequencies of the appearance of letters in an alphabet in certain languages. This is a vast range of fields in which the quantity H is definable, and therefore the SMI is a very useful tool in so many fields of research.

As we shall see in Chapter 3 the entropy is defined only on tiny small sets of probability distributions. Thus, when H is applied to those distributions used in statistical mechanics, it is identical with the statistical mechanical entropy. Therefore, one can safely say that entropy is only one example of the SMI. In other words, the statistical mechanical entropy is a particular case of an SMI, but the SMI, is in general, not the entropy. Unfortunately, confusion of the two concepts abound. The source of this confusion is probably due to von Neumann's suggestion to Shannon to name the quantity H entropy.[5]

In this book, we refer to the quantity defined in Eq. (2.19) as Shannon's measure of information (SMI). We are not interested in the formal proof of the uniqueness of this function. Instead we will survey the properties, and the meanings of the quantity as defined in Eq. (2.19).

2.5.1 *Definition of Shannon's Measure of Information*

In the mathematical theory of information as developed by Shannon, one starts with a random variable X and the corresponding probability distribution $p_1, \ldots, p_n$. Here, we will use a simpler language. We will denote X by an experiment, say throwing a dice, or tossing a coin. The outcomes of this experiment are denoted $A_1, A_2, \ldots, A_n$, and the corresponding probabilities are $p_1, \ldots, p_n$. The set of events $A_1, \ldots, A_n$ is assumed to be *complete*. This means that when we perform an experiment, one and only one of these events must occur. Figure 2.8 shows a board of unit area divided into five mutually exclusive regions and their sum covers the entire board. If we throw a dart on the board it must hit one and only one of the regions denoted $A_1, \ldots, A_n$. The probabilities of these events are supposed to be known and fulfill the condition[6]

$$\sum_{i=1}^{n} p(A_i) = 1 \qquad (2.20)$$

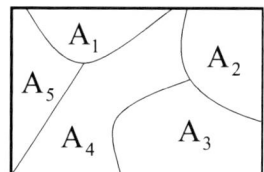

Fig. 2.8 A board of area A is divided into five regions of area A_i such that $A = \sum A_i$.

i.e., the occurrence of either A_1, or A_2 or . . . A_n is certain. The assumption of mutual exclusiveness of each pair of events is written as

$$A_i \cdot A_j = \emptyset \quad \text{for any } i \text{ and } j, \text{ with } i \neq j \qquad (2.21)$$

where $A_i \cdot A_j = \emptyset$ means the intersection of the two events (i.e. both A_i and A_j occurred) is the empty event, the probability of which is zero

$$p(A_i \cdot A_j) = p(\emptyset) = 0, \quad \text{for } i \neq j \qquad (2.22)$$

For instance the experiment of throwing a single dice has six possible outcomes 1, 2, 3, 4, 5, 6. If the dice is well balanced, the probability distribution of this game is $p_1 = p_2 = p_3 = p_4 = p_5 = p_6 = 1/6$. We say that in this case the distribution is *uniform*. On the other hand, the distribution of the events "the dart fell in A_i" as shown in Fig. 2.8 is not uniform.

Qualitatively, *knowing* the distribution is equivalent to having some *information* about the experiment. The question posed by Shannon is how we can measure the "size" of this information.

Before we elaborate on Shannon's measure of information, consider the following simple example. I tossed a coin, looked at the outcome, and you have to *guess* which outcome has occurred. If you guess correctly you win a prize, if not, you get nothing.

Consider the following two cases:

A. I tell you that the coin is fair. This means that the probability distribution is uniform, and the probability of "head" (H), or "tail" (T) are $1/2$, $1/2$.

B. I tell you that the coin is "unfair", it is heavier on one side, and that the probability distribution of the two outcomes is $1/100$ and $99/100$ for the H and the T, respectively.

Intuitively you feel that in case B you are given more information than in case A. More information in what sense?

Suppose that we play the game where I toss a coin and you have to guess the outcome. Clearly, in case A, the chances of guessing the outcome correctly is 50 percent. There is no way you can use the information contained in the uniform distribution to enhance your chances of winning the prize. To put it differently, the uniform distribution does not suggest any "preferred" outcome; given the uniform distribution is tantamount to saying that the distribution conveys *no information* that we can use to win the prize. *If* you play the game A many times, you cannot do any better than guessing the outcomes H or T at random. In this case, you will win the prize, on the average about 50 percent of the time.

On the other hand, if you play game B, you can *use* the *information* contained in the distribution to your advantage. In this particular case, it will be wiser to guess that the outcome was T, rather than H. This does not *guarantee* that you will guess correctly the outcome in each single game. However, if you play the same game (B) many times, and if you always bet on T, you will win on the average about 99 percent of the time. Let us make one generalization of this game. Suppose you are given the distribution $(p, 1 - p)$ for the outcomes H and T, respectively, how would you make the guess? Clearly, if $p = 1, 1 - p = 0$, then you know for *certain* that the result is H. We say that in this case we have the maximum information on the experiment. Suppose now that $p = 9/10$, $1 - p = 1/10$. In this case you do not know the outcome for certain, but it is more *likely* that the result is H rather than T. When $p = 7/10$, $1 - p = 3/10$ again you know that H is more likely, but the degree of certainty is lower (or the degree of uncertainty is higher) than in the previous case. When $p = 1/2$, $1 - p = 1/2$ your degree of certainty is the lowest (or the degree of uncertainty is the highest). Next, suppose that $p = 3/10$, $1 - p = 7/10$. In this case the degree of uncertainty is larger than in the uniform case. You should guess that T is the outcome, and you will win 70 percent of the time. In the case $p = 1/10$, $1 - p = 9/10$, if your guess is T, you will win 90 percent of the time. When $p = 0$, $1 - p = 1$ you know that the result is T, and you can win 100 percent of the time when you guess T. In this case you have been given maximum information.

Thus, when p changes from zero to a half (and $1 - p$ changes from one to a half), the extent of certainty as to the outcome of the experiment, changes from certainty to a minimum (or the uncertainty changes to a maximum). The same is true when p changes from one to half.

Shannon sought a measure of the information contained in a more general distribution, say $p_1, \ldots, p_n$. For instance, a coin is hidden in one of n boxes, Fig. 2.9. We are given the distribution $p_1, \ldots, p_n$, i.e., p_i is the probability that the coin is in box number i.

As we can see from the quotation in Section 2.5, Shannon started with the general concept of information. Information can be subjective or objective, it can be interesting or dull, or even meaningless, and it can be important or totally irrelevant.

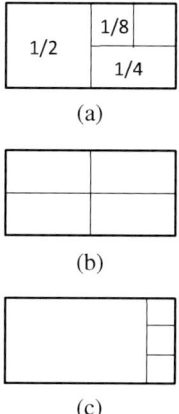

(a)

(b)

(c)

Fig. 2.9 A game with a non-uniform distribution, discussed in Sec. 2.6.

Then he restricted himself to one kind of information; the one contained in a probability distribution. Next, Shannon asked himself: Suppose we have an experiment (or a random variable, or a game), and we know the probability distribution of the outcomes, can we *measure* such information? To answer this question Shannon argued that if such a measure exists then it must have some properties. These are listed in Section 2.5; properties that were deemed to be plausible.

With these plausible requirements of the expected function, Shannon then proved that the only function that satisfies these requirements is of the form

$$H(p_1, \ldots, p_n) = -K \sum_{i=1}^{n} p_i \log p_i \qquad (2.23)$$

where K is some positive constant. In this chapter, we will take $K = 1$, and we will take the base 2 for the logarithm. In Chapter 3 and the rest of the book we will choose K to be the Boltzmann constant, and use the natural logarithm to conform with the units of entropy as are customarily used in the literature. Clearly, the choice of K and the base of the logarithm do not affect the *properties* or the *meaning* of the quantity H. It only affects the units we choose to measure H. In this chapter we use the symbol H as used by Shannon. However, in other chapters we use the acronym SMI instead of H.

2.5.2 Some Elementary Properties of the Function H

We start with the game of tossing a coin with two outcomes. The probability distribution is $(p, 1 - p)$, and the function H for this case is shown in (Fig. 2.7A).

$$H = -p \log_2 p - (1 - p) \log_2 (1 - p) \qquad (2.24)$$

As we can see this function is a continuous function of the single variable p. It also has a maximum at $p = 1/2$. The derivative of H with respect to p is

$$\frac{\mathrm{d}H}{\mathrm{d}p} = -\log_2 p + \log_2(1 - p) \qquad (2.25)$$

The condition for maximum is

$$\frac{\mathrm{d}H}{\mathrm{d}p} = \log_2 \frac{1 - p}{p} = 0 \qquad (2.26)$$

or equivalently

$$p = 1 - p \qquad (2.27)$$

Hence

$$p_{\max} = \frac{1}{2} \qquad (2.28)$$

The value of H at the maximum is

$$H_{\max} = H(p_{\max}) = 1 \qquad (2.29)$$

We can further check the second derivative of H with respect to p, which is

$$(\ln 2)\frac{d^2 H}{dp^2} = \frac{-1}{1-p} - \frac{1}{p} = \frac{-1}{p(1-p)} \le 0 \qquad (2.30)$$

In Eq. (2.30), we use ln for the natural logarithm. Thus, the second derivative is always negative, which means that the function is concave downwards (Fig. 2.7).

Note that the quantity $p \log_2 p$ tends to zero when p tends to zero. This can be easily seen by using L'Hopital's theorem, i.e.

$$\lim_{x \to 0} (x \log x) = \lim_{x \to 0} \frac{\log x}{1/x} = \frac{\frac{d}{dx}(\log x)}{\frac{d}{dx}\left(\frac{1}{x}\right)} = \lim_{x \to 0} \frac{1/x}{-1/x^2} = 0$$

$$(2.31)$$

We now turn to the general case. We have an experiment having n outcomes with probability distribution $p_1, \ldots, p_n$. We shall show that the function H defined in Eq. (2.23) has the following properties.

Continuity of the function H

Since the logarithm function is continuous, the function H is also a continuous function with respect to all the variables $p_1, \ldots, p_n$. It is also a differentiable function of these variables.

Concavity and the maximum of H

The concavity of the function H can be easily proven for the general case. We seek the condition for maximum of the function

$$H = -\sum_{i=1}^{n} p_i \log p_i \qquad (2.32)$$

Subject to the condition

$$\sum_{i=1}^{n} p_i = 1 \qquad (2.33)$$

We use the Lagrange method of undetermined multipliers. We define the auxiliary function

$$F(p_1, \ldots, p_n) = H(p_1, \ldots, p_n) + \lambda \sum_{i=1}^{n} p_i \qquad (2.34)$$

The condition for maximum is

$$(\ln 2)\frac{\partial F}{\partial p_i} = -\log p_i - 1 + \lambda = 0 \quad \text{(for all } p_i) \qquad (2.35)$$

(Here, ln is the natural logarithm and log is the logarithm to the base 2).

Thus, the distribution for which H has a maximum is:

$$p_i^* = 2^{(\lambda-1)} \qquad (2.36)$$

Substituting Eq. (2.36) in Eq. (2.33) we obtain

$$1 = \sum_{i=1}^{n} p_i^* = n2^{(\lambda-1)} \qquad (2.37)$$

Hence the distribution that maximizes H is

$$p_i^* = \frac{1}{n} \quad \text{(for all } i) \tag{2.38}$$

Note that this is a maximum since

$$(\ln 2)\frac{d^2 F}{dp_i^2} = \frac{-1}{p_i} \le 0 \tag{2.39}$$

i.e., the function H is concave downwards.

The value of H for the distribution $p_i^*, \ldots, p_n^*$ is

$$H(p_i^*, \ldots, p_n^*) = -\sum p_i^* \log p_i^*$$

$$= -\sum \frac{1}{n} \log \frac{1}{n} = \log n \tag{2.40}$$

Thus, we find that H has a maximum value when all the p_i are equal, i.e., $p_i^* = 1/n$, and that the value of H at this maximum, $H = \log n$, is a monotonic function of n. This was the second property required by Shannon.

The consistency property of H

The third condition posed by Shannon is sometimes referred to as the *consistency* property of the function H, or the independence on the grouping of the events. This requirement is less obvious intuitively. It essentially states that the amount of information in a given distribution $(p_1, \ldots, p_n)$ is independent of the path, or of the number of steps we choose to acquire this information. In other words, the same *amount* of information is obtained regardless of the way or the number of steps one uses to acquire this information. We shall not need the most general formulation of this property.

Instead, we shall discuss only the case of grouping all the outcomes into two groups. We shall see that such a grouping is essentially equivalent to playing the 20-Q game.

If we divide all the outcomes into two groups, the consistency property of H is very simple. Let us denote the n events by $A_1, A_2, \ldots, A_n$. We need to find which event has occurred knowing that one of these events has occurred. Suppose we group all the n events into two groups, say:

$$(A_1, A_2, A_3, A_4) \quad \text{and} \quad (A_5, A_6, \ldots, A_n)$$

We call the first group G_1, and the second group G_2. The consistency requirement means that the information associated with the original game is the same as the information associated with the new game of G_1 and G_2, plus the average information associated with the groups G_1 and G_2. In terms of the 20-Q game this is equivalent to first finding out which event G_1 or G_2 has occurred, and then find which event has occurred within the group G_1 or G_2. The consistency principle is formulated more generally, and applies to any grouping of the original events into two or more groups.[7] We shall further discuss this meaning of the consistency requirement in Section 2.6.

2.5.3 The Case of an Infinite Number of Outcomes

The case of discrete infinite possibilities is straightforward. First, we note that for a finite and uniform distribution, we had

$$H = \log n \tag{2.41}$$

where n is the number of possibilities. Taking the limit, $n \to \infty$, we get

$$H = \lim_{n \to \infty} \log n = \infty \qquad (2.42)$$

This is clear. If we have to find out one event out of infinite possibilities, we need infinite number of questions.

For a non-uniform distribution, the quantity H might or might not exist, depending on whether the quantity

$$H = -\sum_{i=1}^{\infty} p_i \log p_i \qquad (2.43)$$

converges or diverges.

The case of a continuous distribution is problematic. If we start from the discrete case and proceed to the continuous limit we get into some difficulties.[8] We will not discuss this problem here. Here, we will follow Shannon's treatment for a continuous distribution for which a probability density function $f(x)$ exists. In analogy with the definition of the H function for a discrete probability distribution, we define the quantity H for a continuous distribution. Let $f(x)$ be the density distribution, i.e., $f(x)dx$ is the probability of finding the random variable having values between x and $x + dx$.

We defined the H function as

$$H = -\int_{-\infty}^{\infty} f(x) \log f(x) dx \qquad (2.44)$$

A similar definition applies to an n-dimensional distribution function $f(x_1, \ldots, x_n)$. As in the case of an infinite distribution in Eq. (2.43), the quantity in Eq. (2.44) might be divergent. As we will see in the next chapter, this divergence of the SMI does not cause any problem in actual applications.

2.6 The Various Interpretations of the Quantity *H*

After having defined the quantity H and seen some of its properties, let us discuss a few possible interpretations of this quantity. Originally, Shannon referred to the quantity he was seeking to define as "choice," "uncertainty," "information" and "entropy." Except for the last term which does not have any intuitive meaning, and should not belong to this section, the first three terms do have an intuitive meaning. Let us discuss these with a simple case.

Suppose we are given n boxes and we are told that a coin was hidden in one and only one box. We are also told that the events, "the coin is in box k_1" and "the coin is in box k_2" are mutually exclusive (i.e., the coin cannot be in more than one box), and that the n events form a complete set of events (i.e., the coin is certainly in one of the boxes), and that the box in which the coin was placed was chosen at random, i.e., with probability $1/n$.

The term "choice" is easily understood in the sense that in this particular game, we have to *choose* between n boxes to place the coin. Clearly, for $n = 1$, there is only one box to choose and the amount of "choice" we have is zero; we must place the coin in that box. It is also clear that as n increases, the n is larger, the "choice" we have to select the box in which the coin is to be placed is also larger. The interpretation of H as the amount of "choice," for the case of unequal probabilities is less straightforward. For instance, if the probabilities of say, 10 boxes are $9/10, 1/10, 0, \ldots, 0$, it is clear that we have less choice than in the case of a uniform distribution, but in the general case of unequal probabilities, the "choice" interpretation is not satisfactory. For this reason, we shall not use the "choice" interpretation of H.

The term "Measure of Information" is clearly and intuitively more appealing. If we are asked to find out where the coin is hidden, it is clear, even to the lay person, that we *lack information* on "where the coin is hidden." It is also clear that if $n = 1$, we need no information; we know that the coin is in that box. When n increases, so does the amount of the information we lack, or the missing information. This interpretation can be easily extended to the case of unequal probabilities. Clearly, any non-uniformity in the distribution only increases the information associated with the problem, or decreases the missing information. This reduction in the missing information can be translated into fewer binary questions to be asked. In fact, all the properties of H listed in the previous section are consistent with the interpretation of H as the amount of missing information. For instance, for two sets of independent experiments (or games), the amount of missing information is the sum of missing information of the two experiments. When the two sets are dependent, the occurrence of an event in one experiment affects the probability of the outcome in the second experiment. Hence, having information on one experiment can only reduce the missing information on the second experiment. See also Section 2.7.

The "information" interpretation of H is also intuitively appealing since it is equal to the average number of questions we need to ask in order to *acquire* the missing information. For instance, increasing n will always require more questions to be asked. Furthermore, any deviation from a uniform distribution will reduce the average number of questions.

There are two more interpretations that can be assigned to H which are useful, first, the meaning of H as the amount of *uncertainty*. This interpretation is derived from the meaning of probability. When either $p_i = 1$ or $p_i = 0$, we are *certain* that the event i did occur, or did not occur respectively. If on the other hand, $p_1 = 9/10$, $p_2 = 1/10$, $p_3, \ldots, p_n = 0$, then we have more uncertainty of the outcome compared with the previous example where our uncertainty was zero. It is also intuitively clear that the more uniform the distribution is, the more uncertainty we have about the outcome, and the maximum uncertainty is reached when the distribution of outcomes is exactly uniform. One can also say that H measures the average uncertainty that is removed once we know the outcome. The "uncertainty" interpretation can be applied to all the properties discussed in the previous section.

Perhaps, the simplest interpretation of H is in terms of the 20-Q game. Given n possible outcomes with distribution $p_1, \ldots, p_n$ how difficult is it to play the 20-Q game knowing this distribution? The game we will play is equivalent to the game of hiding a coin in one of n boxes, but here it is easier to *see* what the distribution of outcomes is.

Consider a board of area A divided into n regions, denoted by the numbers $1, 2, \ldots, n$ and areas $A_1, A_2, \ldots, A_n$. I throw a dart on this board at random (imagine that I do it blindfolded). I tell you that the dart has hit some point in the board, and I also tell you the areas of all the regions. Your task is to find out in which of the n areas the dart hit the board by asking only binary questions. The fewer the number of questions you ask the greater will be your prize when you find out where the dart hit.

How will you proceed in asking the questions?

First, you know that the dart hit the board, and all the points on the board that were accessible and were equally likely to be hit. That means that there is probability *one* that the dart is in one of the regions: $1, 2, \ldots, n$.

Furthermore, since all the points on the board are equivalent, it is plausible that the probability of hitting the region i is simply A_i/A. You can take this fraction of areas as your *definition* of the probability that the dart is in the region i. For instance, if there are four areas as in Fig. 2.9a you can estimate that the probability distribution for this game is $(\frac{1}{2}, \frac{1}{4}, \frac{1}{8}, \frac{1}{8})$. We can say that the outcome i (i.e. that the dart had hit the region i) has probability $p_i = A_i/A$.

Clearly, knowing the areas A_i or equivalently the distribution of the outcomes can help you in choosing the strategy of asking questions. Before we discuss the strategy for this particular game consider the two cases b and c in Fig. 2.9.

In case (b) all areas A_i are equal. Hence, each area has probability $p_i = \frac{1}{4}$, i.e. the distribution of outcomes is $(\frac{1}{4}, \frac{1}{4}, \frac{1}{4}, \frac{1}{4})$. In case (c), you are told that the distribution is

$$p_1 = \frac{A_1}{A} = \frac{9}{10}$$

$$p_2 = p_3 = p_4 = \frac{1}{30}$$

According to you, which game is easier to play? If you have any difficulty in answering this question try to play these two games, and "experiment" with different strategies of asking questions.

You will soon find that the best way of playing game b is to divide the four regions into two groups, say (1,2) and (3,4). Once you find out in which *group* the dart is, you will have to ask one more question to find where the dart is. On the other hand, in game (c) you can do better. If you ask: is the dart in region 1? The probability of getting a YES answer is 9/10, and a NO is 1/10.

Clearly, if you play these two games many times you will need on average fewer questions to ask for case (c) than for case (b). Case (a) in Fig. 2.9 is somewhere between (b) and (c).

The SMI for the three games are

(a) SMI $(\frac{1}{2}, \frac{1}{4}, \frac{1}{8}, \frac{1}{8}) = 1.75$

(b) SMI $(\frac{1}{4}, \frac{1}{4}, \frac{1}{4}, \frac{1}{4}) = 2$

(c) SMI $(\frac{9}{10}, \frac{1}{30}, \frac{1}{30}, \frac{1}{30}) = 0.627$

These results fit with what you expected. This is of course not a mathematical proof, only a plausible argument.

There are several other interpretations of the term SMI, or H in Shannon's notation. We will refrain from using the meaning of H as a measure of "entropy," or as a measure of disorder. The first is a misleading term (see Chapter 3) and the second, although commonly used, is very problematic. First, because the terms order and disorder are fuzzy concepts. Order and disorder like beauty and ugliness are very subjective terms and lie in the eyes of the beholder. Second, many examples can be given showing that the amount of information does not correlate with what we perceive as order or disorder.[9]

2.7 Conditional and Mutual Information

In this section we define two important quantities that are derived from the SMI. These are the conditional information and mutual information. These two quantities are very useful in interpreting the SMI of two or more random variables which are not independent. If you are uncomfortable with random variables think of X and Y as two games or two experiments.

Consider first the case of two random variables, X and Y with distributions:

$$p_X(i) = p\{X = x_i\} \quad \text{and} \quad p_Y(j) = p\{Y = y_j\},$$
$$i = 1, 2, \ldots, n, \quad \text{and} \quad j = 1, 2, \ldots, m.$$

$p_X(i)$ is the probability that the outcome x_i has occurred in the experiment X. Likewise, $p_Y(j)$ is the probability that the outcome y_j has occurred in the experiment Y.

We assume for simplicity that $n = m$. Let $p(i, j)$ be the joint probability of occurrence of the events $\{X = x_i\}$ and $\{Y = y_j\}$. The H function defined on the probability distribution $p(i, j)$ is

$$H(X, Y) = -\sum_{i,j} p(i, j) \log p(i, j) \qquad (2.45)$$

The marginal probabilities are defined by

$$p_i = \sum_{j=1}^{n} p(i, j) = p_X(i) \qquad (2.46)$$

and

$$p_j = \sum_{i=1}^{n} p(i,j) = p_Y(j) \tag{2.47}$$

The SMI associated with the random variables X and Y are (all the summations are from one to n).

$$H(X) = -\sum p_X(i) \log p_X(i) \tag{2.48}$$

$$H(Y) = -\sum p_Y(j) \log p_Y(j) \tag{2.49}$$

It can be shown that for any two distributions $\{p_i\}$ and $\{q_i\}$ such that $\sum_{i=1}^{n} q_i = 1$ and $\sum_{i=1}^{n} p_i = 1$ the following inequality holds.[10]

$$H(q_1, \ldots, q_n) = -\sum_{i=1}^{n} q_i \log(q_i) \le -\sum_{i=1}^{n} q_i \log(p_i) \tag{2.50}$$

From Eqs. (2.45), (2.48) and (2.49), we obtain

$$
\begin{aligned}
H(X) + H(Y) &= -\sum_{i} p_X(i) \log p_X(i) - \sum_{j} p_Y(j) \log p_Y(j) \\
&= -\sum_{i,j} p(i,j) \log p_X(i) - \sum_{i,j} p(i,j) \log p_Y(i) \\
&= -\sum_{i,j} p(i,j) \log p_X(i) p_Y(j) \tag{2.51}
\end{aligned}
$$

Applying the inequality in Eq. (2.50) to the two distributions $p(i,j)$ and $p_X(i)p_Y(j)$, we obtain

$$H(X,Y) = -\sum_{i,j} p(i,j)\log p(i,j) \leq -\sum_{i,j} p(i,j)\log p_X(i)p_Y(j)$$

$$= -\sum_{i} p_X(i)\log p_X(i) - \sum_{j} p_Y(j)\log p_Y(j)$$

$$= H(X) + H(Y) \tag{2.52}$$

Thus, we found the important inequality

$$H(X,Y) \leq H(X) + H(Y) \tag{2.53}$$

The equality holds if and only if the two random variables are independent, i.e.

$$p(i,j) = p_X(i)p_Y(i) \quad \text{for } i \neq j \tag{2.54}$$

Hence, in this case

$$H(X,Y) = H(X) + H(Y) \tag{2.55}$$

The last result Eq. (2.55) simply means that if we have two experiments (or two games), the outcomes of which are independent, then the SMI associated with the outcomes of the two experiments is the sum of the SMI associated with the outcomes of each one of the experiment. On the other hand, if the two experiments are dependent then the SMI associated with the compound experiment (X, Y) cannot be larger than the SMI associated with the two experiments separately.

For dependent experiments, we define the conditional probabilities

$$p(y_j|x_i) = \frac{p(x_i \cdot y_j)}{p(x_i)} \tag{2.56}$$

The corresponding conditional quantity is defined as

$$H(Y|x_i) = \sum_j p(y_j|x_i) \log p(y_j|x_i) \tag{2.57}$$

This is simply the SMI associated with the conditional probability defined in Eq. (2.56).

The *conditional SMI of Y given X* is defined as the average of $H(Y|x_i)$, i.e.

$$H(Y|X) = \sum_i p(x_i) H(Y|x_i)$$

$$= -\sum_i p(x_i) \sum_j p(y_j|x_i) \log p(y_j|x_i)$$

$$= -\sum_{i,j} p(x_i \cdot y_j) \log p(x_i \cdot y_j)$$

$$+ \sum_{i,j} p(x_i \cdot y_j) \log p(x_i)$$

$$= H(X, Y) - H(X) \tag{2.58}$$

Thus, $H(Y|X)$ measures the difference between the SMI of X and Y and the SMI of X. This can be rewritten as

$$H(X, Y) = H(X) + H(Y|X)$$

$$= H(Y) + H(X|Y) \tag{2.59}$$

From (2.53) and (2.59) we also obtain the inequality

$$H(Y|X) \leq H(Y) \qquad (2.60)$$

The last inequality means that the SMI of Y can never increase by knowing X. Alternatively, $H(Y|X)$ is the average uncertainty that remains about Y when X is known. This uncertainty is always smaller than the uncertainty about Y. If X and Y are independent, then the equality sign in (2.60) holds. This is another reasonable property that we can expect from a quantity that measures information or uncertainty.

Qualitatively, the meaning of inequality (2.60) is intuitively clear. If we have two independent experiments, performing one does not provide any information about the other. If on the other hand, the two experiments are dependent, then performing one can only *add* information about the other.

Another useful quantity is the *mutual information* defined by

$$I(X; Y) \equiv H(X) + H(Y) - H(X, Y) \qquad (2.61)$$

From the definition of $H(X)$, $H(Y)$ and $H(X, Y)$, we obtain

$$I(X; Y) = -\sum_i p_X(x_i) \log p_X(x_i) - \sum_j p_Y(y_j) \log p_Y(y_j)$$
$$+ \sum_{i,j} p(i,j) \log p(i,j)$$
$$= \sum_{i,j} p(i,j) \log \left[\frac{p(i,j)}{p_X(x_i) p_Y(y_j)} \right]$$
$$= \sum_{i,j} p(i,j) \log[g(i,j)] \geq 0 \qquad (2.62)$$

Note that $I(X; Y)$ is defined symmetrically with respect to X and Y. Sometimes, $I(X; Y)$ is referred to as the average amount of information conveyed by one random variable X, on the random variable Y, and vice versa. Here, $g(i, j)$ [defined in (2.6.2)], is the correlation between the two events $\{X = x_i\}$ and $\{Y = y_j\}$. Thus, we see that $I(X; Y)$ is a measure of an *average logarithm of the correlation* (in the sense of $\log[g(i, j)]$), i.e. it is a measure of the extent of *dependence* between X and Y. The inequality 2.62 follows from the inequality 2.53. The equality holds for independent X and Y. It should be noted that the correlation between two specific events, $g(i, j)$ could be larger or smaller than one, i.e., $\log[g(i, j)]$ could be negative or positive. However, the *average* of $\log[g(i, j)]$ is always positive, i.e. the mutual information is always positive. It is zero when the two experiments are independent. Thus, $I(X; Y)$ measures the average reduction in SMI associated with the experiment X resulting from knowing Y, and vice versa. The relationship between $H(X), H(Y), H(X, Y)$ and $I(X; Y)$ is shown in the diagram of Fig. 2.10.

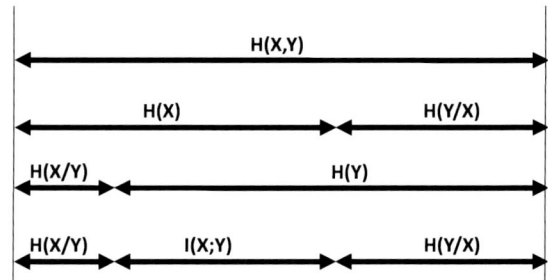

Fig. 2.10 The relationship between the quantities $H(X), H(Y)$, $H(X, Y)$ and $I(X; Y)$.

2.8 Concluding Remarks

The content of this chapter is not directly relevant to statistical thermodynamics. However, I urge the reader to get some degree of familiarity with the concept of SMI. There are two reasons for doing so. First, the SMI is a very general, useful and interesting quantity. It is used in many fields of science. Therefore, there is a good chance that one day you will encounter a problem where the SMI may be useful in measuring or interpreting certain quantities associated with an experiment.

Second, the SMI is indispensable for understanding entropy. This is more than just one application of SMI. It is an application that makes the concept of entropy simple, clear and mystery-free. Without this interpretation the entropy remains highly mysterious and a hard to comprehend concept. We shall see in the next chapter that the entropy is a special case of the SMI.

3

Application of SMI to Systems of Non-Interacting Particles

In the previous chapter we introduced the quantity referred to as the Shannon measure of information (SMI). We discussed the definition of this quantity, its properties and its interpretations without any reference to thermodynamics. We mentioned that Shannon himself renamed his quantity "entropy," and many scientists still refer to SMI as the entropy. This practice is unfortunate and should be avoided.

The origin of this practice is found in Tribus' story[1]

"What's in a name? In the case of Shannon's measure, the naming was not accidental. In 1961 one of us (Tribus) asked Shannon what he had thought about when he had finally confirmed his famous measure. Shannon replied: "My greatest concern was what to call it. I thought of calling it 'information,' but the word was overly used, so I decided to call it 'uncertainty.' When I discussed it with John von Neumann, he had a better idea. Von Neumann told me, "You should call it entropy, for two reasons. In the first place your uncertainty function has been used in statistical mechanics under that name.

In the second place, and more important, no one knows what entropy really is, so in a debate you will always have the advantage."

The SMI of tossing a coin or throwing a dice has nothing to do with the thermodynamic entropy. However, as we will see in this chapter the thermodynamic entropy may be derived from SMI. In this sense, entropy is a special case of SMI. As such entropy bears the same properties as well as the same interpretations as SMI.

In this chapter we will derive an explicit expression for SMI of two ideal systems; a mono-atomic, classical gas, and a system of N adsorbed particles. Each system contains N indistinguishable and non-interacting (ideal) particles. The microscopic state of an ideal gas is described by the locations and the momenta (classical description), of all its particles. It is assumed that all the energy of the system consists of only the translational kinetic energies of the particles (sometimes these are referred to as "structure-less" particles), which is approximately true for mono-atomic gases. We will see that the SMI of an ideal gas is essentially the equation derived independently by Sackur and Tetrode in 1912, based on Boltzmann's definition of entropy.

The second system is much simpler than an ideal gas. This is a system of N particles adsorbed on M sites. Assuming no interactions between the particles, we derive the SMI of this system, which turns out to be the same as the entropy of the adsorbed particles. From this entropy we can also obtain the Langmuir adsorption isotherm.

3.1 The SMI of an Ideal Gas

In this section we apply the SMI to an ideal gas. We will see that SMI of an ideal gas, is up to a multiplicative constant equal to the entropy of the gas.

3.1.1 *The Locational SMI of an Ideal Gas*

Consider the case of one particle free to move in a one-*dimensional* "box" of length L. Clearly, there are infinite points where the center of the particle may be found. However, we are never interested in the *exact* point in which the particle is, but rather in which interval dx it is, where dx is very small (Fig. 3.1). The formal problem we pose is to find the maximum value of the SMI

$$H(location\ in\ 1D) = -\int_0^L f(x) \log[f(x)]dx \qquad (3.1)$$

subject to the constraints

$$\int_0^L f(x)dx = 1 \qquad (3.2)$$

where $f(x)dx$ is the probability of finding the particle in an interval dx at x. The function $f(x)$ that maximizes H is

$$f^*(x) = \frac{1}{L} \qquad (3.3)$$

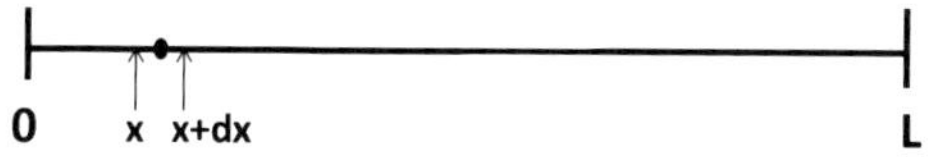

Fig. 3.1 A particle in 1D "box" of length L. We are interested in the probability of finding the center of particle between x and $x + dx$.

and the corresponding SMI[13]

$$H_{\max}(\textit{location of one particle along the x-axis}) = \log L \quad (3.4)$$

We shall refer to the density function $f^*(x)$ as either the function that maximizes H, or as the *equilibrium density function*. The latter interpretation will be clear when we interpret the final result of H as the entropy of an ideal gas at equilibrium.

In this chapter we use the logarithm to the base 2. It should be noted that the results (3.3) and (3.4) were obtained for the function H as defined in Eq. (3.1). In practice we always use a discrete division of the segment L into a finite number of cells. However, in this and in the next section we will use the continuous language for both the locations and the velocities (or the momenta). In Section 3.1.4, we will apply the quantum mechanical uncertainty principle, which imposes a discretization on the joint space of locations and momenta. We note here that $\log L$ is to be understood as a measure of the missing information about the location of the particle in the 1D system of length L. The larger the L the larger the SMI. One should keep in mind that the SMI is actually a divergent quantity. However, in all the application of Eq. (3.4), we either make a *discretization* of the length L, or take the *difference* of the SMI for the two values of L. In both cases we remove any problem arising from the divergence of the integral in (3.1).[2]

The generalization of (3.3) and (3.4) to the three dimensional case is straightforward. Suppose the particle is confined to a cubic box of edge L and volume $V = L^3$. Clearly, the SMI associated with the y-axis and the z-axis will be the same as in

Eq. (3.4). Furthermore, we assume that the events: "being at some location x", "being at some location y", and "being at some location z" are independent events. Therefore, the SMI associated with the location x, y, z within the cube of volume V is the sum of the SMI associated with the three axes. Thus, if we use the short hand notation $H_{\max}(x)$ for the quantity in (3.4) we can write

$$H_{\max}(x, y, z) = H_{\max}(x) + H_{\max}(y) + H_{\max}(z)$$
$$= 3 \log L = \log V \qquad (3.5)$$

and the equilibrium density is

$$f^*(x, y, z) = \frac{1}{L} \times \frac{1}{L} \times \frac{1}{L} = \frac{1}{L^3} = \frac{1}{V} \qquad (3.6)$$

We next extend the result (3.5) to the case of N *independent* and *distinguishable* (D) particles. We also use the short hand notation: $R_i = \{x_i, y_i, z_i\}$ for the locational vector of particle i, Fig. 3.2, and, $R^N = \{R_1, R_2, \ldots R_N\}$ for the locational vector of all the N particles.

Since the particles are *independent*, the SMI of the N particles is simply the sum of the SMI of all the single particles, and since the SMI for a single particle (3.5) is the same for each

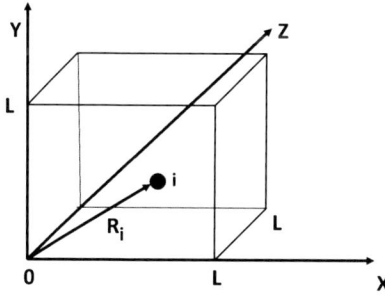

Fig. 3.2 The locational vector R_i of the ith particle.

particle we have for the N, independent and distinguishable particles.

$$H_{\max}^{D}(\boldsymbol{R}^{N}) = N H_{\max}(x, y, z) = N \log V \qquad (3.7)$$

Note that we added the superscript D for *distinguishable* particles. We shall see in the next section that the fact that the particles are *indistinguishable* (ID) introduces a correlation between the particles which causes a *reduction* in the SMI of the N particles. Note also that we still retain the subscript "max." This will be important when we shall identify the maximal value of H with the entropy of the system at equilibrium.

3.1.2 *The Mutual Information due to the Indistinguishability of the Particles*

Consider two particles, denoted 1 and 2, distributed in M equivalent cells. The particles are independent and distinguishable (D). The number of possible arrangements of one particle is simply M and the corresponding SMI is:

$$H(1) = \log M \qquad (3.8)$$

The same is true for the second particle, i.e.

$$H(2) = \log M \qquad (3.9)$$

Therefore, the SMI of the two particles, denoted $H(1, 2)$ is the sum of $H(1)$ and $H(2)$

$$H^{D}(1, 2) = H(1) + H(2) = 2 \log M \qquad (3.10)$$

Thus, the number of possible arrangements of two independent and distinguishable particles in M cells is simply M^{2}

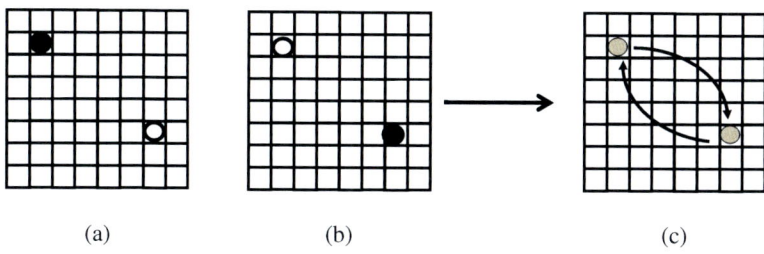

Fig. 3.3 Two different configurations (a) and (b) of two distinguishable particles become a single configuration (c) when the particles are indistinguishable.

and the corresponding SMI is $H^D(1, 2)$ given in (3.10). Note that the particles do not interact. In this example there is no limit on the number of particles that can occupy the same cell.

If the particles are indistinguishable (ID) then the counting of the number of arrangements is different. The reason is quite simple. We do not distinguish between two configurations that are obtained by exchanging the particles. Figure 3.3 shows two possible configurations (a) and (b) of two particles in the cells. Clearly, when the two particles are distinguishable, these two configurations are *different*, and therefore should be counted as two configurations. However, if the two particles are indistinguishable, this amounts to erasing the colors or the labels of the two particles; hence, the two configurations coalesce into one (c). Thus, in general the total number of configurations is *reduced* when we "un-label" the particles.

In the general case of N indistinguishable particles in M cells the counting of the total number of configurations of N distinguishable particles is simply M^N. The number of configurations of N indistinguishable particles is more complicated to count. However, for the case that $N \ll M$ (i.e., when the number of cells is so large that the occurrence of more than one particle in a cell is a rare event), the number of configurations is reduced from M^N to $M^N/N!$

Exercise: Calculate the total number of configurations of $N = 2$ in $M = 4$ cells, when the particles are distinguishable and when they are indistinguishable. Keep $N = 2$ fixed and change M to 10, 100, 1000 and extrapolate to very large M.

In the example shown in Fig. 3.3 when the number M is very large, the number of configurations of the system of the two indistinguishable particles is $M/2$, the corresponding SMI is

$$H^{ID}(1, 2) = \log \left(\frac{M^2}{2} \right) \tag{3.11}$$

The difference between $H^D(1, 2)$ and $H^{ID}(1, 2)$ can be interpreted in terms of *mutual* information (see Section 2.7), i.e.

$$H^{ID}(1, 2) = H(1) + H(2) - I(1; 2) \tag{3.12}$$

where

$$I(1; 2) = \log 2 > 0 \tag{3.13}$$

In the more general case of N *indistinguishable* particles on M cells, such that $N \ll M$, we have

$$H^D(1, 2, \ldots, N) = \sum_{i=1}^{N} H(i) = \log M^N \tag{3.14}$$

$$H^{ID}(1,2,\ldots,N) = H^{D}(1,2,\ldots,N) - \log N!$$
$$= \log(M^{N}/N!) \tag{3.15}$$
$$I(1;2;,\ldots;N) = \log N! \tag{3.16}$$

We conclude that because the particles are indistinguishable, there is a *correlation* between the particles, which causes a *reduction* of the SMI. We have calculated the mutual information of indistinguishable particles, by calculating the change in the number of configurations of the system of distinguishable particles, caused by "erasing the labels" on the particles. One can also give a probabilistic interpretation of the mutual information, in terms of correlation between events. We shall not need this interpretation here. The reader should be convinced by checking a few examples that whenever we "un-label" the particles, the number of configurations or arrangements is always reduced, and as a result of which the value of the SMI of the system is reduced too.

3.1.3 *The Momentum SMI*

We start again with particles moving along the 1D system of length L. We are interested in the probability density of finding a specific particle with velocity between v_x and $v_x + dv_x$. We assume that the particles can have any value of v_x from $-\infty$ to $+\infty$, but we require that the average kinetic energy of the particles is constant.

The formal mathematical problem is to find the maximum value of the SMI

$$H(\text{momentum in 1D}) = \int_{-\infty}^{\infty} f(v_x) \log[f(v_x)] dv_x \tag{3.17}$$

subject to the two constraints

$$\int_{-\infty}^{\infty} f(v_x)dv_x = 1 \qquad (3.18)$$

$$\int_{-\infty}^{\infty} \frac{mv_x^2}{2} f(v_x)dv_x = \frac{m\langle v_x^2 \rangle}{2} = \frac{m\sigma^2}{2} \qquad (3.19)$$

where σ^2 is the variance of the distribution $f(v_x)$, and m is the mass of a single particle.

The solution to this problem is [3]

$$f^*(v_x) = \frac{\exp[-v_x^2/2\sigma^2]}{\sqrt{2\pi\sigma^2}} \qquad (3.20)$$

Before we proceed to calculate the distribution of *speeds* in three dimensions we first express the variance σ^2 in terms of the absolute temperature T.

The average kinetic energy of a particle moving in one dimensional system at equilibrium is:

$$\frac{m\langle v_x^2 \rangle}{2} = \int_{-\infty}^{\infty} \frac{mv_x^2}{2} \frac{\exp[-v_x^2/2\sigma^2]}{\sqrt{2\pi\sigma^2}} dv_x = \frac{m\sigma^2}{2} \qquad (3.21)$$

The average kinetic energy of a particle moving in three dimensions is given by

$$\frac{m\langle v^2 \rangle}{2} = \frac{m\langle v_x^2 \rangle}{2} + \frac{m\langle v_y^2 \rangle}{2} + \frac{m\langle v_z^2 \rangle}{2} = 3\frac{m\langle v_x^2 \rangle}{2} \qquad (3.22)$$

where v is the *speed* of the particles. From the kinetic theory of gases we have the relation between the absolute temperature and the average kinetic energy[3]

$$k_B T = \frac{2}{3} \frac{m \langle v^2 \rangle}{2} \qquad (3.23)$$

From Eqs. (3.21), (3.22) and (3.23) we identify σ^2 as

$$\sigma^2 = \frac{k_B T}{m} \qquad (3.24)$$

with this identification we rewrite the distribution of velocities (3.20) as

$$f^*(v_x) = \sqrt{\frac{m}{2\pi k_B T}} \exp[-m v_x^2 / 2 k_B T] \qquad (3.25)$$

Figure 3.4 shows the distribution $f^*(v_x)$ for various values of T (for this illustration we take $m = 1$, $k_B = 1$). We see that the larger the temperature, the larger the *spread* of the

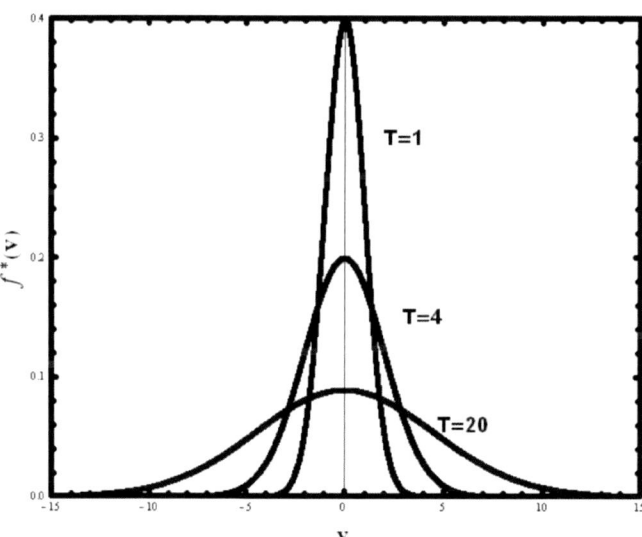

Fig. 3.4   The velocity distribution in 1D at different temperatures. The curves were calculated from Eq. (3.25) for $m = 1$, $k_B = 1$.

distribution of the velocities. Thus, the average "width" of the distribution is a measure of either the variance σ^2 or of the temperature T.

In terms of the temperature we rewrite the SMI associated with the equilibrium distribution $f^*(v_x)$ as

$$H_{\max}(v_x) = \frac{1}{2} \log \left(2\pi \, e \, k_B T / m \right) \qquad (3.26)$$

The meaning of this quantity is derived from the meaning of the SMI. Note again that the SMI of any continuous variable has a divergent part. However, in actual application of $H_{\max}$ we either make a *discretization* of the infinite range $(-\infty, \infty)$ into a finite number of intervals, or we take *differences* in the values of H between two states; here, two temperatures. The important result we have obtained is that the larger the temperature (or equivalently the average kinetic energy of the particles) the larger is the SMI or the uncertainty associated with the distribution of the velocities.

We next assume that the velocities along the three axes v_x, v_y, v_z are independent. Therefore, the SMI for a single particle moving with velocities v_x, v_y, v_z is

$$H_{\max}(v_x, v_y, v_z) = H_{\max}(v_x) + H_{\max}(v_y) + H_{\max}(v_z)$$

$$= 3H_{\max}(v_x) = \frac{3}{2} \log \left(2\pi \, e \, k_B T / m \right)$$

$$(3.27)$$

For the purpose of constructing the entropy of an ideal gas we shall need the distribution of the momenta. This is simply obtained by the transformation $p_x = mv_x, p_y = mv_y,$

$p_z = mv_z$. Hence, the one dimensional distribution of momenta is

$$f^*(p_x) = \frac{\exp[-p_x^2/2mk_BT]}{\sqrt{2\pi mk_BT}} \qquad (3.28)$$

The corresponding SMI in one dimension is

$$H_{\max}(p_x) = \frac{1}{2}\log(2\pi emk_BT) \qquad (3.29)$$

And in three dimensions

$$H_{\max}(p_x, p_y, p_z) = \frac{3}{2}\log(2\pi emk_BT) \qquad (3.30)$$

For a system of N independent particles the SMI is simply the sum of the SMI of each particle, hence we have

$$H_{\max}(\boldsymbol{p}^N) = \frac{3N}{2}\log(2\pi emk_BT) \qquad (3.31)$$

where $\boldsymbol{p}^N = (\boldsymbol{p}_1, \ldots, \boldsymbol{p}_N)$

We now have the SMI associated with the momentum of the N particles. At this stage we also derive the distribution of *speeds*. Although we shall not need this distribution, it is helpful to have an idea of the form of this distribution.

The speed of the particles is defined by the positive square root

$$v = \sqrt{v_x^2 + v_y^2 + v_z^2} \qquad (3.32)$$

This is the absolute value of the speed of a particle having the components of velocities v_x, v_y and v_z along the three axes.

Since the motions along the three axes are independent we have

$$f^*(v_x, v_y, v_z) = f^*(v_x)f^*(v_y)f^*(v_z)$$

$$= \left(\frac{m}{2\pi k_B T}\right)^{3/2} \exp\left[\frac{-m(v_x^2 + v_y^2 + v_z^2)}{2k_B T}\right]$$

$$= \left(\frac{m}{2\pi k_B T}\right)^{3/2} \exp\left[\frac{-mv^2}{2k_B T}\right] \qquad (3.33)$$

$f^*(v_x, v_y, v_z)dv_x dv_y dv_z$ is the probability of finding a particle with velocities between v_x and $v_x + dv_x$, v_y and $v_y + dv_y$ and v_z and $v_z + dv_z$. The *speed* v defined in (3.32) can be obtained with infinitely many combinations of v_x, v_y and v_z.

The distribution of the speeds is obtained by transforming to spherical polar coordinates and integrating over all the angles. The result is

$$f^*(v) = \left(\frac{m}{2\pi k_B T}\right)^{3/2} 4\pi v^2 \exp\left[\frac{-mv^2}{2k_B T}\right] \qquad (3.34)$$

Here, $f^*(v)dv$ is the probability of finding a particle with a *speed* between v and $v + dv$. Note carefully the difference between the distribution of velocities (3.25), and the distribution of speeds (3.34). The velocity v_x can be either positive or negative, its distribution is Normal (Fig. 3.4), and centered at $v_x = 0$. It is a symmetric distribution, i.e. the particle has the same probability of moving with a velocity v_x as with $-v_x$. On the other hand, the *speed* distribution is not

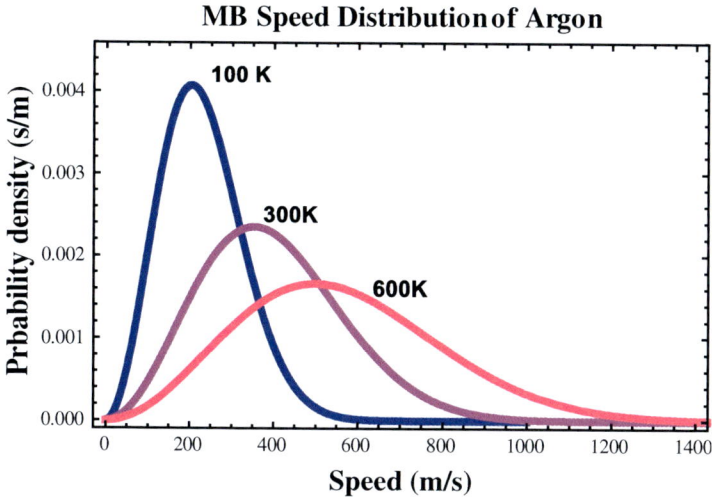

Fig. 3.5   Distribution of speeds at different temperatures.

symmetric; the average and the most probable speeds are not identical, as shown in Fig. 3.5.

3.1.4 The Mutual Information Associated with the Uncertainty Principle

In Section 3.1.1, we saw that the continuous SMI as defined in Eq. (3.1) for the location of a single particle within the "box" of length L is

$$H_{\max}(x) = \log L \tag{3.35}$$

We also note that if we are interested only in the location of the particle in one of the cells of length $h = L/n$, Fig. 3.6, we need not use (3.35), but the discrete analog of (3.35) which is obtained from $H_{\max}(x)$ by subtracting $\log h$, i.e., instead of (3.35) we write

$$H_{\max}(x) = \log L - \log h = \log n \tag{3.36}$$

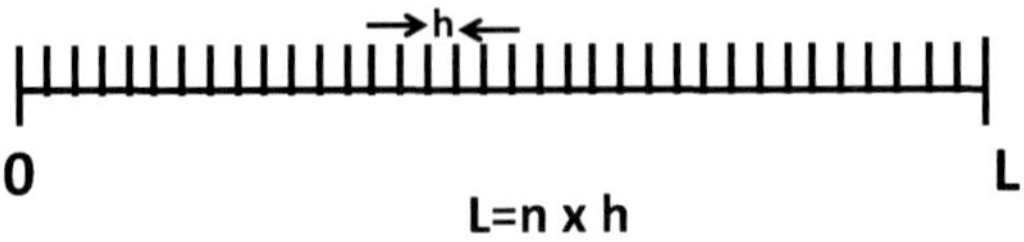

Fig. 3.6 Discretization of the segment $(0, L)$, into n cells $L = n \times h$.

In Section 3.1.1, we calculated the locational SMI. In Section 3.1.3 we calculated the momentum SMI. We now want to find out the SMI associated with *both* the location and the momentum of the particles.

Classical thinking would have led us to conclude that the SMI associated with both the location *and* momentum of a particle should be the sum of the two SMIs. However, quantum mechanics dictates that the two experiments of determining the location and the momentum of a particle are *not independent* events. This is the well-known Heisenberg uncertainty principle. For our case the uncertainty principle states that we cannot determine *both* the location and the momentum within an accuracy of the order of h, here h is the Planck constant $h = 6.626 \times 10^{-34} Js$. It follows that the SMI of a particle is not the sum of the SMI associated with its location and the SMI associated with its momentum, but a corrected sum, after taking into account the uncertainty principle.

This is the same situation we encountered when passing from the continuous segment $(0, L)$ to the discrete number of cells in Fig. 3.6. Here, we also divide the entire space of locations and momenta into cells of size h, where now h is the Planck constant.

In terms of SMI we can say that there is a *correlation* between the location and the velocity (or momentum) of a particle. This correlation can be cast in the form of *mutual information*, i.e.

$$I \ (uncertainty \ principle) = \log h > 0 \qquad (3.37)$$

or equivalently

$$H_{\max}(x, p_x) = H_{\max}(x) + H_{\max}(p_x) - \log h \qquad (3.38)$$

This is the SMI for one particle in one dimension. For the three dimensional case we have

$$H_{\max}(x, y, z, p_x, p_y, p_z) = H_{\max}(x, y, z) + H_{\max}(p_x, p_y, p_z)$$
$$- 3 \log h \qquad (3.39)$$

i.e., we subtract $\log h$ for each degree of freedom.

Finally, for N indistinguishable and non-interacting particles, we have

$$H^{ID}(1, 2, \ldots, N) = H^{D}_{\max}(\boldsymbol{R}^{N}) + H^{D}_{\max}(\boldsymbol{p}^{N})$$
$$- \log N! - 3N \log h \qquad (3.40)$$

This is an important result. To obtain the SMI of N particles, described by their locations and momenta, we first treat the particles as being distinguishable and classical. In this case we can sum the SMI associated with all the locations $(\boldsymbol{R}^{N})$, and all the momenta $(\boldsymbol{p}^{N})$ of the particles. Then, we correct by deducting two kinds of mutual information due to the fact that the particles are not classical. One, because the particles are indistinguishable, and two, due to the Heisenberg uncertainty principle. These two corrections *reduce* the total SMI by the amount $\log N! + 3N \log h$.

3.2 The SMI and the Entropy of an Ideal Gas

In the previous section, we have calculated the maximal value of the SMI of a system of N non-interacting and indistinguishable particles.

Recall that the SMI may be defined for *any* distribution. It can be defined for any distribution of locations and any distribution of momenta, not necessarily at equilibrium. It can be characterized for any number of particles and can be defined for distinguishable or indistinguishable particles. All these have nothing to do with entropy. Up to this point you can rightfully regard the SMI as a quantity that measures the size of the 20-Q games (Where is the particle? What is its velocity or momentum?)

In this section, we will be interested in *very special distributions*. These are the distributions of locations and momenta that maximized the corresponding SMI. We denoted these special distributions by f^* or p^*, and the corresponding SMI by $H_{\max}$. However, it is known that starting with any arbitrary distribution of locations and momenta, the system will tend to a limiting equilibrium distribution; the uniform distribution for locations, and the Normal distribution for the momenta.

In this section we take a huge conceptual leap, from SMI of games to a fundamental concept of thermodynamics. As we will soon see, this leap is rendered possible by recognizing that the SMI associated with the equilibrium distribution of locations and momenta of a large number of indistinguishable particles is *identical* (up to a multiplicative constant) with the statistical mechanical entropy of an ideal gas. Since the statistical mechanical entropy of an ideal gas has the same properties as the thermodynamic entropy as defined by Clausius we can declare that this special SMI is *identical* to the entropy of an ideal gas. This is a very remarkable achievement. Recall that von Neumann suggested calling "entropy," the SMI. This was unfortunate. It caused a great confusion in the field. In general, the SMI has nothing to do with the entropy. Only when one applies the SMI to a *special* distribution does it become identical to entropy.

We recall that the SMI of a system of N particles at equilibrium has two contributions due to location and momentum, and two corrections, one due to the fact that the particles are indistinguishable, and the second due to the uncertainty principle. Thus, we have the following expression for the SMI of N non-interacting particles at equilibrium.

$$H^{ID}(1, 2, \ldots, N)$$

$$= H_{\max}(locations) + H_{\max}(momenta)$$

$$-I \; (uncertainty \; principle) - I \; (indistinguishability)$$

$$= [N \log V] + \left[\frac{3N}{2} \log (2\pi e m k_B T) \right]$$

$$-[3N \log h] - [\log N!]$$

$$= N \log \left[\frac{V}{N} \left(\frac{2\pi m k_B T}{h^2} \right)^{3/2} \right] + \frac{5N}{2} \qquad (3.41)$$

This expression is identical with the equation obtained by Sackur and Tetrode in 1912 based on Boltzmann's definition of entropy.

Exercise: Repeat the procedure to arrive at Eq. (3.41) for the case of a mixture of c-components ideal gas.[4]

We derived Eq. (3.41) from considerations based on SMI corresponding to the locations and the momenta of the particles. We also added two corrections due to the uncertainty principle and to the fact that the particles are indistinguishable, in the form of mutual information.

In order to obtain the expression for the entropy of an *ideal* gas all we have to do is to use the natural logarithm and multiply H^{ID} by the Boltzmann's constant k_B

$$S = (k_B \ln 2)H \qquad (3.42)$$

The multiplication by a constant k_B, as well as the usage of the natural logarithm determines the units in which we measure the entropy. It does *not affect* the *meaning* of *entropy*, i.e. as a SMI associated with location and momenta of a system of N non-interacting particles at equilibrium. We apply this identity between entropy and SMI for a thermodynamic system, i.e., when N is very large, V very large but the density N/V is constant.

We have assumed throughout that the particles do not interact with each other. Normally, absence of interactions is considered to be equivalent to independence. However, as we have seen in this chapter, *dependence* can occur even between non-interacting particles. When the particles interact with each other, a new kind of dependence is introduced. This dependence introduces an additional correlation between the particles, which causes another reduction in the SMI. We will discuss many systems with interacting particles in the rest of this book. See also Ben-Naim (2008).

From now on we shall *define* the entropy of an ideal gas by Eq. (3.42). The *meaning* of the entropy is derived from the four terms listed in Eq. (3.41) i.e., we have the SMI associated with the locations of the particles, the momenta of the particles and the two corrections due to the uncertainty principle and because the particles are indistinguishable.

3.3 Fundamental Properties of the Entropy Function $S(E,V,N)$

In the last section we derived the function $H^{ID}(1, 2, \ldots, N)$ which, when applied to a system of N non-interaction particles in a volume V and at temperature T, is identified with the *entropy* of the system. We write this function as

$$S(T, V, N) = Nk_B \ln \left[\frac{V}{N} \left(\frac{2\pi mk_B T}{h^2} \right)^{3/2} \right] + \frac{5Nk_B}{2}$$

$$(3.43)$$

However, for reasons discussed below the fundamental function for the entropy is not $S(T, V, N)$, but the function $S(E, V, N)$ where E is the total energy of the particles. For an ideal monatomic gas, the energy of the system is simply the total kinetic energy of the particles given by

$$E = N \frac{m\langle v^2 \rangle}{2} = \frac{3}{2} Nk_B T \qquad (3.44)$$

Eliminating T from Eq. (3.44) and substituting in Eq. (3.43) we obtain the *fundamental* function

$$S(E, V, N) = Nk_B \ln \left[\left(\frac{V}{N} \right) \left(\frac{E}{N} \right)^{3/2} \right]$$

$$+ \frac{3k_B N}{2} \left[\frac{5}{3} + \ln \left(\frac{4\pi m}{3h^2} \right) \right] \quad (3.45)$$

This function is fundamental for two reasons:

1. For any thermodynamic system of N particles, in a volume V and having total energy E, the entropy function $S(E, V, N)$ provides all the thermodynamic quantities of that system.
2. For any thermodynamic system of N particles in a volume V and having total energy E, the entropy function $S(E, V, N)$ has a maximal value at equilibrium.

Here are some clarifying comments:

(i) The N particles in the system can be atoms or molecules, they can have translational, vibrational, rotational, etc. energies. If there are c-components, we interpret N as the vector $N_1, \ldots, N_c$ where N_i is the number of molecules (or moles) of species i.

(ii) The volume of the system is defined by the boundaries of the specific thermodynamic system under study. We always assume that the system is macroscopic, i.e. its dimensions are very large compared to the molecular diameters of the particles. We also assume that surface effects are negligible. Also, we assume that there are no *external* fields that operate on the system.

(iii) The total energy of the system E, includes all the internal energies of the atoms and molecules, as well as the potential energies of interactions. Actually, the internal energy of the system is always defined with respect to some arbitrary chosen zero. The First Law of Thermodynamics is essentially a statement on the changes in the internal energy of the system caused by exchange of either heat or work with its surroundings.

(iv) Maximum with respect to what? This is a tricky question. A great deal of confusion exists in the literature regarding the answer to this question.

In calculus, when we say that a function $y = f(x)$ has a single maximum, we mean that there is a value of x such that the value of y at this x is maximal compared to all the values of y obtained for any other value of x. A function can have more than one maximum, in which case one needs to define each maximum locally. The mathematical requirements for a maximum of the function $y = f(x)$ at the point x^* are

$$\left. \frac{df(x)}{dx} \right|_{x=x^*} = 0 \qquad (3.46)$$

and

$$\left. \frac{d^2 f(x)}{dx^2} \right|_{x=x^*} < 0 \qquad (3.47)$$

Equation (3.46) means that the *slope* of the function at $x = x^*$ is zero, and Eq. (3.47) means that the *curvature* is negative.

By generalization, if we are given a function $w = f(x, y, z)$ and we say that there exists a maximum at some point

(x^*, y^*, z^*) we mean that w has the largest value when x, y and z are varied in the neighborhood of x^*, y^*, z^*. Similar conditions on the derivatives of $f(x, y, z)$ with respect to each of the variable x, y and z as in (3.46) and (3.47) apply to this maximum.

In some formulations of the Second Law of Thermodynamics it is stated that the entropy of a system tends to a maximum at equilibrium. Such a statement is faulty in two respects: First, it does not specify which *function* has a maximum, and second, it does not specify with respect to which *variable* the entropy has a maximum. Without such specification the general statement that "entropy always reaches a maximum" is not valid.

In thermodynamics, we assume that we have the liberty of choosing, at will, the independent variables. For instance, for a one-component system we may choose the independent variables E, V, N or T, V, N or T, P, N, etc., For each of these independent variables, we have different entropy function; $S(E, V, N)$, $S(T, V, N)$ and $S(T, P, N)$, etc.

Clearly, there are many possible choices of sets of independent variable, and the corresponding entropy functions. The Second Law of Thermodynamics, when applied to the entropy, applies *only* to the specific *entropy function* $S(E, V, N)$. That is the reason why this particular function is referred to as the *fundamental entropy function*.

Now for the question: "Maximum with respect to which variable?"

In the mathematical examples given above when we have written a function $f(x)$, we knew that $f(x)$ has a maximum with respect to the variable x. In the case of

$f(x, y, z)$ the maximum is with respect to variation of the independent variables x, y and z. In general, it is understood implicitly that the function has a maximum with respect to the *arguments* of the function, i.e., x in $f(x)$ or x, y, z in $f(x, y, z)$. In thermodynamics we must first choose the independent variables to describe the system, say E, V, N or T, V, N, etc. Second, unlike in the mathematical practice, the entropy function *does not* have a maximum with respect to the independent variables E, V, N. On the contrary, these variables must be kept *constant*. What we vary to obtain the maximum of $S(E, V, N)$ is some *internal distribution* while keeping E, V, N constants.

Figure 3.7 shows that the function $S(E, V, N)$ is actually a monotonically increasing function of E, V and N. Furthermore, the three curves are concave downward. This is an important attribute of the dependence of the entropy on the variables E, V and N.[5]

In this chapter, we have calculated the maximum of SMI with respect to all possible *distributions* of the locations and momenta of the particles. The Second Law applies to the function $S(E, V, N; \ distribution)$. Here, we add the new

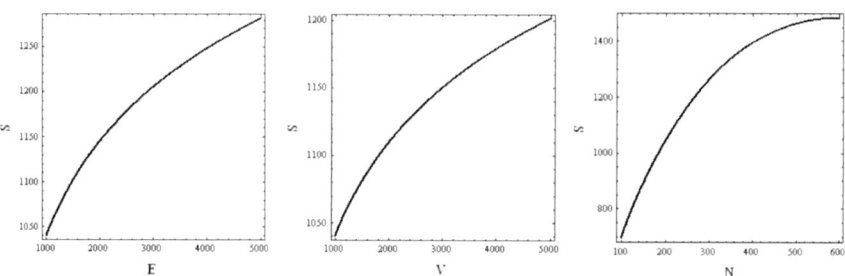

Fig. 3.7 The general form of the entropy function $S(E, V, N)$, with respect to E, V and N.

argument "distribution" to the function *S(E, V, N)*. Failing to identify the argument with respect to which the entropy is maximal leads to serious errors.[6]

Note that the Second Law is valid for any system: ideal gas as well as non-ideal gas. It is valid, whether or not we have the explicit function *S(E, V, N)*. The important thing is that the variables E, V, N are *kept constant*. Such a system is referred to as an *isolated system*, i.e. a system that does not interact with its surroundings; and it does not exchange energy, volume or matter with its surroundings.

We stress again that the Law of Maximal Entropy is valid only when the entropy is viewed as a function of E, V, N, and it has a maximum with respect to variable "distributions." It is not true for the entropy functions $S(T, V, N;$ distribution), or $S(T, P, N;$ distribution), etc. In practical applications of the Second Law we formulate it in terms of either the Gibbs energy function $G(T, P, N)$ or the Helmholtz energy$(A(T, V, N))$.

3.4 Some Concluding Remarks Regarding the Relationship between Entropy and SMI

This chapter introduced the concept of entropy of an ideal gas as a special case of SMI. It is important to understand the conceptual steps that led us from SMI to entropy.

First, we start with the general concept of information. We do not have a definition of information, but we certainly know what information is. Shannon sought out a measure of information, and he found one; not for *any* information, but for certain kinds of information i.e. those associated with

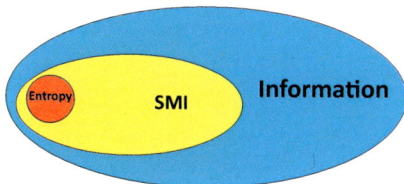

Fig. 3.8 Schematic relationship between the "Entropy," SMI and the general concept of "Information" (see text).

well-defined distributions. We have referred to this measure as Shannon's measure of information (SMI). Clearly, the kind of information on which we can define the SMI is only a subset of vast, in fact, infinite set of information. Figure 3.8 schematically shows the relationship between the various kinds of information. The outer region denoted "Information" is not a well-defined region at all. All we want to convey in this figure is that the SMI is defined on a very small subset of "information." It is by no means clear whether or not any "information" can be measured, and certainly not clear, if the SMI can be applied to the general information. Thus, at this stage we have "distilled" a concept of information on which we can apply SMI.

It should be noted that points in the subset denoted by SMI in Fig. 3.8 are not SMI, but all the possible *distributions* on which the SMI can be defined. It is intuitively clear that this subset is very small compared with all possible "information"; yet, this subset is still very large. It contains infinite number of possible distributions. For example, the distribution pertaining to tossing a coin, throwing a dice, measuring the volume of an expanding gas, or the frequency of occurrence of a letter in an alphabet of a specific language.

Within this large subset, denoted SMI, we identify an even smaller subset denoted "Entropy" in Fig. 3.8. This subset contains *all* the *distributions relevant* to a *thermodynamic system* at *equilibrium*. Clearly, these are only a tiny set of distributions, compared to all possible distributions on which the SMI is definable.

There is a subtle point in formulating the principles of maximum entropy in thermodynamics, and the maximum SMI in all other fields. In thermodynamics, the maximum of entropy is over the *manifold* of constrained *equilibrium states*. In all other cases, and especially in the cases discussed in this chapter, we use the principle of maximum SMI over *all* possible distributions. But it is only the *maximal* SMI which we identify as entropy. For all distributions, except those that maximize SMI, the system is not at equilibrium. Therefore, for such systems the entropy is not even defined.

It turns out that the entropy as derived in this chapter is up to an additive constant and pending a choice of the appropriate units, identical to the thermodynamic entropy introduced by Clausius. Conceptually, SMI and Entropy are very different from each other. However, both are measures of the information contained, or associated with a distribution. In this sense we can safely say that entropy is a particular case of SMI. On the other hand, one cannot refer to any SMI as entropy. This terminology is unfortunately very much in use, and was adopted by Shannon himself.

Once you have grasped the logical steps and the sequence of concepts leading from "general information" to SMI, to entropy, you do not have to memorize the derivation of the entropy. Be impressed and perhaps astonished by the finding that entropy, a concept that was spawned out of considerations about heat engines, turned out to be nothing but a certain measure of information. This link between two apparently very different concepts is not less amazing than the connection between temperature and kinetic energy of motion of the particles.

The reader should realize that our procedure to obtain the equation for the entropy of an ideal gas differs in a fundamental way from the Sackur–Tetrode method.

Sackur and Tetrode estimated the *number of states* Ω of an ideal gas contained in a volume V. Once you have Ω you can calculate the entropy by using the Boltzmann relationship (Eq. 2.1), $S = k_B \ln \Omega$. This procedure provides the correct entropy of an ideal gas, but does not offer any *interpretation* of the entropy. In our approach we did not calculate Ω, then the entropy. Instead we directly calculated the SMI of an ideal gas (with respect to locational and momentum distribution) at equilibrium. We then showed that this SMI is identical (up to a constant) to the entropy of an ideal gas. With this identification, the entropy we obtained carries exactly the same meaning and the same interpretation as the SMI.

We hope that you feel comfortable with the new concept of entropy. Disregard the original meaning of this word, just remember what an SMI is. You should remember also that entropy is defined, at least in thermodynamics, only for equilibrium states. Furthermore, for thermodynamic systems at fixed values of the energy E, volume V and number of particles N, entropy attains a maximum value over all possible *distributions*. These distributions are relevant to equilibrium states. Otherwise, the entropy is not even defined.

3.5 The SMI of *N* Ligands Adsorbed on *M* Sites

We conclude this chapter with one more example of a system for which the SMI can be easily calculated.

This is a simpler example than the case of an ideal gas, in two senses: First, the particles do not have velocities (or momenta), and second, the possible location of each particle is a *finite set* of M sites, not a continuum. This model is very useful in describing binding phenomenon which we shall describe in Chapter 5.

The model is very simple: N simple particles (i.e. having no internal degrees of freedom) are distributed on M sites (Fig. 3.9). We normally assume that $N \ll M$. Each site can accommodate at most one particle. The particle at the site is motionless, and there are no interactions between particles on adjacent sites.

For N distinguishable (D) particles, the total number of configurations is

$$W^D = M(M-1)(M-2)\cdots M-(N-1)$$
$$= \frac{M!}{(M-N)!} \tag{3.48}$$

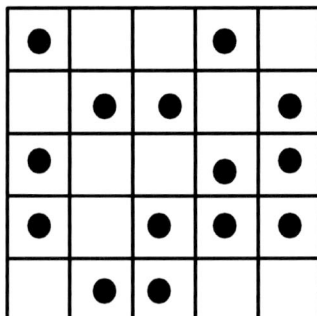

Fig. 3.9 A surface with $N = 14$ particles distributed over $M = 25$ sites.

There are M possible locations to place the first particle, $(M - 1)$ for the second, and so on.

If we assume that all W^D configurations are equally probable, then we can say that the SMI of such a system is

$$SMI^D = \log W^D \tag{3.49}$$

If the particles are indistinguishable (ID), then exchanging any two particles on different sites does not produce a new configuration. Hence, the number of configurations is

$$W^{ID} = \frac{M!}{(M - N)!N!} \tag{3.50}$$

and the corresponding SMI is

$$SMI^{ID} = \log W^{ID} \tag{3.51}$$

We could have viewed the factor $N!$ as a correction to the counting of the configurations of the ID particles, by first assuming that the particles are distinguishable (D), then introduce a correlation between the particles due to the fact that the particles are indistinguishable. In either case Eq. (3.51) gives the SMI of N indistinguishable particles distributed on M sites.

For very large N, and M, with $N \ll M$, we may rewrite Eq. (3.51) using the Stirling approximation as follows

$$SMI^{ID} = \log \frac{M!}{(M - N)!N!} = M \log M$$

$$-N \log N - (M - N) \log (M - N) \tag{3.52}$$

Denote by $\theta = N/M$, the fraction of occupied sites, and by $1 - \theta = (M - N)/M$ the fraction of unoccupied sites, we can rewrite (3.52) as

$$SMI^{ID} = M[-\theta \log \theta - (1 - \theta) \log (1 - \theta)]$$

$$= M \ SMI \ (one \ site) \qquad (3.53)$$

We now recognize that $(\theta, 1 - \theta)$ is a probability distribution for one site. Picking up a site at random, the probability that it is occupied is θ, and that it is unoccupied is $(1 - \theta)$. Therefore, the expression in the squared brackets is Eq. (3.53) is the SMI per single site which we denote by SMI (one site). Thus, the SMI of the entire system is viewed in Eq. (3.53) as M times the SMI of a single site.

Note carefully that in Eq. (3.51) we viewed all the configurations W^{ID} as having equal probabilities. The corresponding SMI was calculated in Eq. (3.51) for this uniform distribution. On the other hand, the distribution $(\theta, 1 - \theta)$ for each site is non-uniform, and the corresponding SMI was calculated from Shannon's measure for the non-uniform distribution. The multiplication by M in Eq. (3.53) is a result of the independence of the M sites, each having the same SMI.

In Chapter 5, we shall see that the SMI in Eq. (3.53) is related to the entropy of the system of adsorbed ligands on independent sites. From this relationship we shall derive the Langmuir adsorption isotherm.

Part II

Applications

4

Thermodynamics of Ideal Gases

Experiments with gases were carried out long before the discovery and the acceptance of the atomic nature of matter. The most outstanding laws that bear the names of Boyle, Charles and Gay-Lussak are now referred to as the equation of state of ideal gas. The Boyles' Law, (Fig. 4.1) states that at a given temperature, the product of the pressure P, and the volume V of the gas is a constant. The Charles' Law states that at constant pressure, the volume of the gas is a linear function of the temperature. These two empirical laws may be combined into one equation of state of an ideal gas;

$$PV = nRT \qquad (4.1)$$

where n is the number of moles and R is referred to as the *gas constant*. In this chapter we shall study the molecular theory of ideal gases. We shall obtain quite a few useful and informative results, one of which is the equation of state of an ideal gas. An ideal gas is one of the simplest systems than can be solved exactly with the methods of classical statistical mechanics.

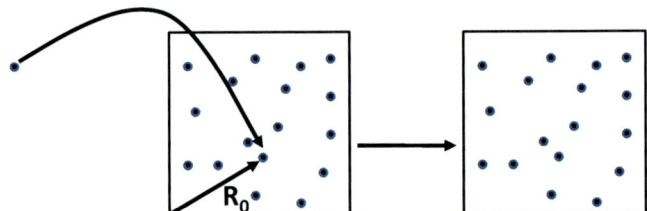

Fig. 4.1 Adding one particle to a system at T, P, N. First the particle is placed at a fixed position.

In Chapter 1, we derived the partition function (PF) of a classical system based on analogy with the quantum mechanical PF. Here, we shall apply this classical PF for the case of ideal gases. It should be noted however, that the term "classical PF" is not exactly classical, since two corrections have been introduced in the classical analog of the PF, which originate from quantum mechanical considerations.

4.1 Thermodynamics of a One-Component Ideal Gas

A theoretical *ideal gas* is defined by the requirement that there are no intermolecular interactions among the particles, i.e.

$$U(X^N) \equiv 0 \qquad (4.2)$$

for any configuration $X^N \equiv (X_1, X_2, \ldots, X_N)$. Where X_i is the vector comprising the location (R_i) and the orientation (Ω_i) of the particle i.

A theoretical ideal gas is only a *model* for a system with very weak intermolecular interactions. In any real gas, there are always some interactions between the particles. However, at very low densities, the average distance between any two particles is very large. At such distances, the total intermolecular interaction-energy among all the particles is negligible compared with the kinetic energies of the particles. Hence, a *real* gas at very low density behaves as an ideal gas.

From the classical canonical partition function (PF), derived in Chapter 1 we have for an ideal gas

$$
Q(T, V, N) = \frac{q^N}{(8\pi^2)^N \Lambda^{3N} N!} \int \cdots \int dX^N
$$

$$
= \frac{q^N}{(8\pi^2)^N \Lambda^{3N} N!} \left[\int_V dR \int_0^{2\pi} d\phi \right.
$$

$$
\left. \times \int_0^\pi \sin\theta d\theta \int_0^{2\pi} d\psi \right]^N
$$

$$
= \frac{q^N V^N}{\Lambda^{3N} N!} \tag{4.3}
$$

For simple spherical particles, sometimes referred to as "structure-less" particles, Eq. (4.3) reduces to

$$
Q(T, V, N) = V^N / \Lambda^{3N} N! \tag{4.4}
$$

Here, q is the internal PF, and Λ^3 is the momentum PF.

The Helmholtz energy of the system is related to $Q(T, V, N)$ by Eq. (1.14).

Note that q and Λ^3 depends on the temperature but not on the volume V, nor on N. An important consequence of this is that the equation of state of an ideal gas is independent of the particular molecules constituting the system. To see this, we derive the expression for the pressure (see Section 1.4), by differentiating (4.4) with respect to the volume:

$$P = k_B T \left(\frac{\partial \ln Q}{\partial V}\right)_{T,N} = \frac{k_B TN}{V} = \rho k_B T \qquad (4.5)$$

where $\rho = N/V$ is the number density.

In statistical mechanics we use the Boltzmann constant k_B which is related to the gas constant R by $R = k_B N_{av}$, where N_{av} is the Avogadro number.

This is an important result. First, note that this equation of state is *universal*; it does not depend on the properties of the molecules. This behavior is not shared by other thermodynamic quantities of the ideal gas.

For instance, the chemical potential is obtained by differentiation of (4.4) with respect to N and using the Stirling approximation, we get

$$\mu = -k_B T \left(\frac{\partial \ln Q}{\partial N}\right)_{T,V} = k_B T \ln (\Lambda^3 q^{-1}) + k_B T \ln \rho$$

$$= \mu^{0g}(T) + k_B T \ln \rho \qquad (4.6)$$

where $\rho = N/V$ is the number density and $\mu^{0g}(T)$ is referred to as the standard chemical potential as defined in Eq. (4.6). The latter conveys the properties of the individual molecules in the system. Note that the value of $\mu^{0g}(T)$ depends on the choice of units of ρ. The quantity $\rho \Lambda^3$, however, is dimensionless. Hence, μ is independent of the choice of the concentration units.

The entropy of an ideal gas may be obtained from (4.6) by differentiation with respect to T.

$$S = k_B \ln Q + k_B T \left(\frac{\partial \ln Q}{\partial T} \right)_{V,N} = \frac{5}{2} k_B N$$

$$-Nk_B \ln (\rho \Lambda^3 q^{-1}) + k_B T N \frac{\partial \ln q}{\partial T} \qquad (4.7)$$

For simple particles ($q = 1$) this reduces to:

$$S = \frac{5}{2} k_B N - N k_B \ln \rho \Lambda^3 \qquad (4.8)$$

At this stage pause and compare the result (4.8) with a similar result we obtained in Chapter 3 (Eq. 3.41). First, we rewrite the expression for entropy in the form

$$S = \frac{5}{2} k_B N + k_B N \ln \left[\frac{V}{N} \left(\frac{2\pi k_B m T}{h^2} \right)^{3/2} \right] \qquad (4.9)$$

We see that except for the Boltzmann constant k_B, and the difference in the base of the logarithm, (4.9) is identical with the expression (3.41) for the SMI of an ideal gas. This is a remarkable result. The two expressions were derived from very different considerations. The first, (3.41) was derived from Shannon's measure of information applied to the distribution of locations and momentum of non-interacting and structure-less particles. The second (4.9) was derived from the PF of a classical system, based on the formalism of statistical mechanics which is based on the Boltzmann definition of the entropy. Therefore, we can conclude that the entropy of an ideal gas is (up to a multiplicative constant) identical with the SMI when applied to an ideal gas. Thus, we can say that the entropy of an ideal gas is a special case of an SMI. From this identification we can draw another important conclusion. The *meaning* of the entropy of an ideal gas is the same as the meaning of the SMI as we have discussed in Chapter 2. One can also show that the meaning of the entropy of a system of interacting particles is also the same as the meaning of SMI. We shall not discuss this aspect of the entropy of a non-ideal gas. The interested reader is referred to Ben-Naim (2008).

The energy of a system of simple particles is obtained from Eqs (1.14), (4.4) and (4.8), i.e.

$$E = A + TS = k_B TN \ln \rho \Lambda^3 - k_B TN$$

$$+ T \left(\frac{5}{2} k_B N - N k_B \ln \rho \Lambda^3 \right) = \frac{3}{2} k_B TN \quad (4.10)$$

Thus, the energy in this case is entirely due to the kinetic energy of the particles. $3k_B T/2$ is the average kinetic energy of one particle.

The heat capacity for a system of simple particles is obtained directly from (4.10) as

$$C_V = (\partial E / \partial T)_V = \frac{3}{2} k_B N \quad (4.11)$$

which may be viewed as originating from the accumulation of $\frac{1}{2} k_B$ for each translational degree of freedom of a particle. For molecules having also rotational degrees of freedom, we get

$$C_V = 3 k_B N \quad (4.12)$$

which is built up of $\frac{3}{2}k_B N$ from the translational and $\frac{3}{2}k_B N$ from the rotational degrees of freedom. If other internal degrees of freedom are present, there are additional contributions to C_V.

In all of the aforementioned discussions, we left unspecified the internal partition function of a single molecule denoted q. This, in general includes contributions from the rotational, vibrational, and electronic states of the molecule. Assuming that these degrees of freedom are independent, the corresponding internal partition function (PF) may be factored into a product of the PFs for each degree of freedom, namely

$$q(T) = q_r(T)q_v(T)q_e(T) \qquad (4.13)$$

We shall never need to use the explicit form of the internal PF in this book. However, such knowledge is needed for the actual calculation, for instance, of the equilibrium constant of a chemical reaction.

Each of the factors in (4.13) has the general form

$$q(T) = \sum_i \omega_i \exp[-\beta\varepsilon_i] \qquad (4.14)$$

where ε_i is the ith energy level, and ω_i is the corresponding degeneracy.

It is important to emphasize that the calculation of the internal partition function of a molecule is possible only when we are given the energy levels and the corresponding degeneracies. These are presumed to be provided either from experiments or from quantum mechanics. Statistical mechanics is not involved in calculating these quantities.

Note also that each of the factors in (4.13) is a function of T, but not a function of either V or N. The internal PF, $q(T)$ is a property of a single molecule and it does not depend on the presence of other molecules in the volume V. In general, when the density of the gas $\rho = N/V$ is large, the energy levels of the molecule might be affected by the interactions between the molecules, in which case the whole factorization of q from the configurational PF in (4.3), or the factorization of $q(T)$ in (4.13) might not be valid.

We quote here the general form of the three factors in (4.13)

For the electronic PF we write

$$q_e(T) = \sum_i \omega_i^{(e)} \exp[-\beta \varepsilon_i^e] \qquad (4.15)$$

where $\varepsilon_i^{(e)}$ is the ith electronic energy level and $\omega_i^{(e)}$ is the corresponding degeneracy. In some cases the electronic excitation energy from the ground state is very large compared with $k_B T$. In such a case it is sufficient to retain only the first term on the rhs of (4.15), namely

$$q_e(T) = \omega_0^{(e)} \exp\left(-\beta \varepsilon_0^e\right)$$
$$\times \left\{ 1 + \frac{\omega_1^{(e)}}{\omega_0^{(e)}} \exp[-\beta(\varepsilon_1^e - \varepsilon_0^e)] + \cdots \right\} \qquad (4.16)$$

Since $\beta(\varepsilon_i^e - \varepsilon_0^e)$ are very large and positive ($i > 0$) at room temperature, we can neglect all the terms except the first one on the rhs of (4.16) to obtain

$$q_e(T) = \omega_0^{(e)} \exp\left(-\beta \varepsilon_0^e\right) \qquad (4.17)$$

For the vibrational degree of freedom, the energy levels, as calculated by quantum mechanics for the harmonic oscillator are given by

$$\varepsilon_n = \left(n + \frac{1}{2}\right)h\upsilon \quad n = 0, 1, 2, 3 \ldots, \tag{4.18}$$

In (4.18), υ is the classical frequency of a one-dimensional harmonic oscillator with a reduced mass μ and force constant f:

$$\upsilon = \frac{1}{2\pi}\sqrt{f/\mu} \tag{4.19}$$

The vibrational PF is thus

$$q_{\upsilon}(T) = \sum_{n=0}^{n=\infty} \exp\left(-\beta\varepsilon_n\right) = \exp\left(-\beta h\upsilon/2\right)$$

$$= \sum_{n=0}^{\infty} \exp\left(-\beta nh\upsilon\right) = \frac{\exp\left(-\beta h\upsilon/2\right)}{1 - \exp\left(-\beta h\upsilon\right)} \tag{4.20}$$

The rotational partition function of an assymmetric diatomic molecule is derived from the energy levels of a linear rigid rotator. These are also calculated by quantum mechanics:

$$\varepsilon_n = \frac{n(n+1)h^2}{8\pi^2 I} \quad n = 0, 1, 2, 3, \ldots, \tag{4.21}$$

The degeneracy of the nth energy level is

$$\omega_n = 2n + 1 \tag{4.22}$$

and I is the moment of inertia about the center of mass of the molecule. Thus,

$$q_r(T) = \sum_{n=0}^{\infty} (2n + 1) \exp[-\beta n(n + 1)h^2/8\pi^2 I]$$

$$\approx \int_0^{\infty} (2n + 1) \exp[-\beta n(n + 1)h^2/8\pi^2 I] dn$$

$$= \frac{8\pi^2 I k_B T}{h^2} \tag{4.23}$$

The passage from the sum to the integral is valid whenever

$$\frac{h^2}{8\pi^2 I k_B} \ll T \tag{4.24}$$

For a symmetrical diatomic molecule, the rotational PF is

$$q_r(T) = \frac{8\pi^2 I k_B T}{2h^2} \tag{4.25}$$

The factor 2 in the denominator arises because each pair of configurations resulting from the exchange of the two nuclei is considered as a single configuration. In an assymmetric molecule such a pair produces two distinguishable configurations.

For a general molecule, the assymmetric top, the rotational PF is

$$q_r(T) = \frac{\pi^{1/2}}{\sigma} \left(\frac{8\pi^2 I_A k_B T}{h^2} \right)^{1/2} \left(\frac{8\pi^2 I_B k_B T}{h^2} \right)^{1/2}$$

$$\times \left(\frac{8\pi^2 I_C k_B T}{h^2} \right)^{1/2} \tag{4.26}$$

where I_A, I_B and I_C are the three principal moments of inertia of the molecule and σ is the symmetry number. This is introduced, as in the case of the symmetrical linear molecule, to account for the number of indistinguishable configurations obtained by the rotation of the molecule.

We note again that in all the applications in this book, the internal partition functions are presumed to be known.

4.2 The Chemical Potential of an Ideal Gas

In Section 4.1 we derived the thermodynamic quantities of an ideal gas. Of particular importance is the chemical potential, Eq. (4.6) which we rewrite

$$\mu = k_B T \ln q^{-1} + k_B T \ln \frac{N}{V} \Lambda^3 \qquad (4.27)$$

Here, q is the internal partition function of a single molecule, V the volume of the system, N the number of particles, and Λ^3 the momentum partition function.

It is convenient to assign meaning to the two terms in Eq. (4.27) as follows: The chemical potential is defined in the T, V, N ensemble as

$$\mu = \left(\frac{\partial A}{\partial N} \right)_{T,V} = A(T, V, N + 1) - A(T, V, N) \qquad (4.28)$$

where the second equality follows from the extensive property of the Helmholtz energy and the fact that the addition of one particle to a macroscopic system can be viewed as an infinitesimal change in the variable N.

Suppose that instead of adding one particle to the system, we place the particle at a fixed position in the system, say at R_0. The corresponding change in the Helmholtz energy is defined by

$$\mu^* = A(T, V, N + 1, R_0) - A(T, V, N)$$

$$= -k_B T \ln \left[\frac{q^{N+1} V^N}{\Lambda^{3N} N!} \right]$$

$$+ k_B T \ln \left[\frac{q^N V^N}{\Lambda^{3N} N!} \right] = k_B T \ln q^{-1} \quad (4.29)$$

We see that this quantity is equal to the first term on the rhs of Eq. (4.27). We shall refer to μ^* as the *pseudo chemical potential*

Combining Eqs. (4.27) and (4.29), we obtain

$$\mu = \mu^* + k_B T \ln \frac{N}{V} \Lambda^3 \quad (4.30)$$

Thus, the work required to introduce one particle into the system is divided into two parts. The first part is the work required to place the particle at a fixed position in the system (see Fig. 4.1). Next, the particle is freed; the corresponding work is $k_B T \ln \rho \Lambda^3$. We shall refer to this term as the *liberation Helmholtz energy*. It is easy to show that the corresponding *liberation Gibbs energy* has the same form as in (4.30) with the replacement of V by the average volume in the T, P, N ensemble.

When the particle is released from its fixed position, there are three factors contributing to the change in the Helmholtz energy. First, it can acquire translational kinetic energy, the corresponding contribution being $k_B T \ln \Lambda^3$.

Second, the particle can now wander in the entire volume V, the corresponding contribution being $-k_B T \ln V$. Finally, when the particle is at a fixed position, it is distinguishable from all the N indistinguishable particles. This is the reason for having $N!$ in both terms on the rhs of (4.29). Once the particle is released, it is no longer distinguishable from other members of the same species. We shall refer to the term $k_B T \ln N$ as being due to the *assimilation* of one particle by N indistinguishable particles. Together the three contributions give rise to the *liberation* term $k_B T \ln \rho \Lambda^3$, where $\rho \Lambda^3$ is a dimensionless quantity. Note that since we are using classical PF, $\rho \Lambda^3 < 1$, the liberation Helmholtz energy is always negative.

Each of the factors mentioned above can be changed independently. If we change the temperature (at V, N constant), the change in the chemical potential is

$$\mu(T_2) - \mu(T_1) = \mu^*(T_2) - \mu^*(T_1) + k_B T_2 \ln[\Lambda^3(T_2)]$$
$$-k_B T_1 \ln[\Lambda^3(T_1)] \qquad (4.31)$$

On the other hand, at a fixed temperature we may change the volume, say from V to $2V$; the corresponding change of the chemical potential is

$$\mu(2V) - \mu(V) = k_B T \ln \frac{V}{2V} = k_B T \ln \frac{1}{2} \qquad (4.32)$$

The change in the chemical potential for the expansion process is negative. Finally, we can change N and keep T and V constant, for instance, by eliminating $\frac{1}{2}N$ of the particles.

The corresponding change in the chemical potential is

$$\mu\left(\frac{1}{2}N\right) - \mu(N) = k_B T \ln \frac{N}{2N} = k_B T \ln \frac{1}{2} \quad (4.33)$$

Formally, in both Eqs. (4.32) and (4.33), the density $\rho = N/V$ has been cut to half its initial value. However, the change in density in (4.32) is due to expansion, whereas in (4.33) it is due to a change in the number of particles.

Exercise: Calculate the change in the entropy of an ideal gas upon expansion from V to $2V$ in an isolated system. Try to interpret this change in entropy based on SMI of an ideal gas.

4.3 Mixture of Ideal Gases

The study of mixtures of ideal gases is important for two reasons: First, these systems serve as a reference for the study of real mixtures and solutions; and second, these are special cases of systems in which a chemical reaction occurs. As we will see in Section 4.4, a system of reacting species in an ideal-gas mixture reduces to a non-reacting ideal gas mixture whenever we block the flow of material from one species to another, e.g. by adding an inhibitor or by removing a catalyst.

In this section we consider a system consisting of two components of structureless particles of species A and B. Let N_A and N_B be the number of particles of species A and B, respectively, contained in a volume V at temperature T. The T, V, N_A, N_B, PF of this system is

$$Q(T, V, N_A, N_B) = \frac{V^{N_A+N_B}}{\Lambda_A^{3N_A} \Lambda_B^{3N_B} N_A! N_B!} \quad (4.34)$$

As expected, the PF of the system is simply the product of the PFs of the two pure systems of A and B each in the same volume V at temperature T.

From Eq. (4.34) it follows that the Helmholtz energy of the system is

$$A(T, V, N_A, N_B) = -k_B T \ln Q(T, V, N_A, N_B)$$

$$= -k_B T \ln Q(T, V, N_A)$$

$$- k_B T \ln Q(T, V, N_B)$$

$$= A(T, V, N_A) + A(T, V, N_B) \quad (4.35)$$

The Helmholtz energy of the system is simply the sum of the Helmholtz energies of the two pure components, each occupying separately a volume V.

From Eq. (4.35) we can derive all the other thermodynamic quantities of the system.

For instance, the entropy of the system is

$$S(T, V, N_A, N_B) = -\left(\frac{\partial A(T, V, N_A, N_B)}{\partial T}\right)_{V, N_A, N_B}$$

$$= S(T, V, N_A) + S(T, V, N_B) \quad (4.36)$$

and similarly the energy of the system is

$$E(T, V, N_A, N_B) = A(T, V, N_A, N_B) + TS(T, V, N_A, N_B)$$

$$= E(T, V, N_A) + E(T, V, N_B) \quad (4.37)$$

From these results we can conclude that a system of two (non-interacting) components characterized by the variables T, V, N_A, N_B is thermodynamically equivalent to *two*

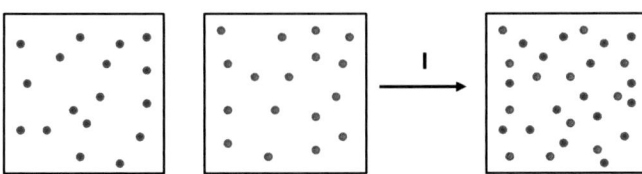

Fig. 4.2 A process of mixing two ideal gases with no change in energy and in entropy.

separate systems, each containing one component in the same volume V and temperature T. In other words, when we combine two systems characterized by T, V, N_A and T, V, N_B into a single system characterized by T, V, N_A, N_B there is no change in any of the thermodynamic quantities. Fig. 4.2 shows a process of mixing of two components with no change in entropy. This conclusion is not valid for systems of interacting or reacting particles.

In particular we note that when the two systems are combined (Fig. 4.2) the species A and B do mix in the process, yet no entropy change is observed. We can therefore conclude that the process of *mixing* (of ideal gases) by itself has no effect on the entropy of the system. The reason for this result is simple. In the process of mixing in Fig. 4.2, there is no change in either the accessible volume, or the velocity distribution. Also, the number of particles of each species is unchanged. Therefore, we can expect no change in the Shannon's measure of information, as well as in the entropy when the two systems are combined.

Exercise: Generalize Eqs. (4.36) and (4.37) for any number of non-interacting species.

There is another process in which mixing is involved and in which both the Helmholtz energy and the entropy do

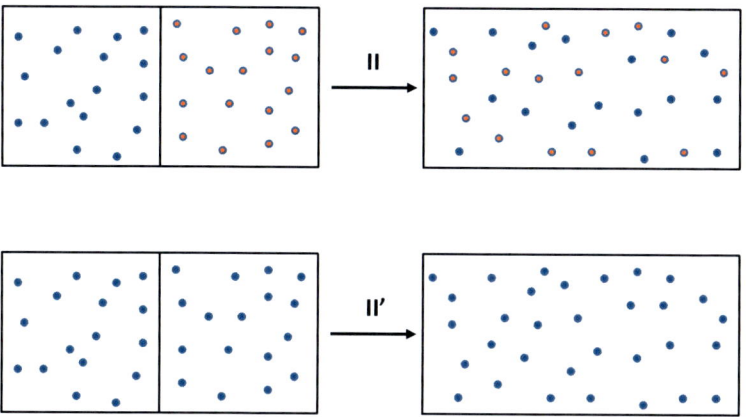

Fig. 4.3A (II') Mixing to two different ideal gases A and B. (II) "Mixing" of the same gas.

change. Consider process II in Fig. 4.3. We start with two pure components A and B as in the previous experiment. We let the two components mix in a volume $2V$, instead of V as in the previous experiment. This can be achieved by removing a partition separating the two compartments.

The partition function in the initial state is exactly the same as in Eq. (4.34). In the final state however, we have

$$Q(T, 2V, N_A, N_B) = \frac{(2V)^{N_A + N_B}}{\Lambda_A^{3N_A} \Lambda_B^{3N_B} N_A! N_B!} \qquad (4.38)$$

The change in the Helmholtz energy in this process is:

$$\Delta A = A(T, 2V, N_A, N_B) - A(T, V, N_A) - A(T, V, N_B)$$

$$= -k_B T (N_A + N_B) \ln (2V) + k_B T N_A \ln V$$

$$+ k_B T N_B \ln V = -k_B T N_A \ln 2 - k_B T N_B \ln 2 \qquad (4.39)$$

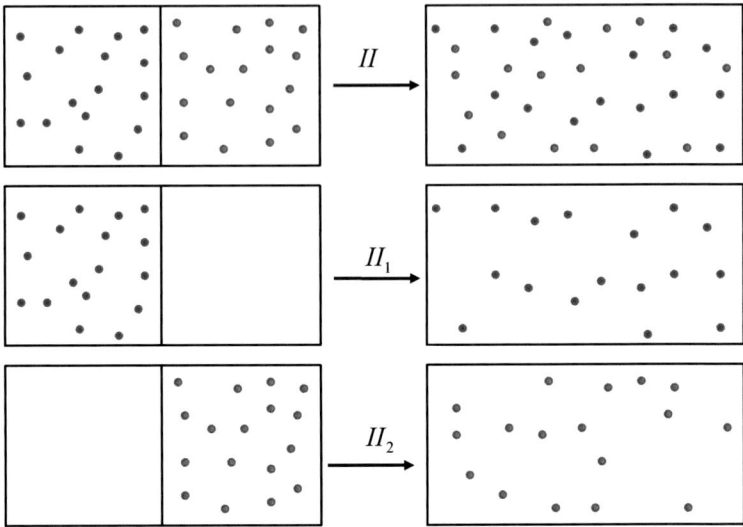

Fig. 4.3B (II) Mixing to two different ideal gases A and B. and the equivalent expansion of two pure gases ($II_1 + II_2$).

This is sometimes referred to as the "free energy of mixing."

The corresponding entropy change is

$$\Delta S = k_B N_A \ln 2 + k_B N_B \ln 2 \qquad (4.40)$$

which is often referred to as the "entropy of mixing." However, Eq. (4.39) shows that this term originates not from the *mixing* process but from the *expansion* process; each gas expands from V to $2V$. The same entropy change would have been obtained in the process of expanding each component *separately* from V to $2V$, in Fig. 4.3B ($II_1 + II_2$).

From the thermodynamic point of view, the two processes depicted in Fig. 4.3B (*II* and ($II_1 + II_2$)) are equivalent, producing the same entropy change as in Eq. (4.40). Nevertheless, ΔS of one process is referred to as the "entropy of mixing," whereas the same entropy change in the second process is referred to as the "entropy of expansion." It is true that in the first process we *observe*, mixing as we observed "mixing" in the process shown in Fig. 4.2. The additional complexity of process in Fig. 4.3 (*II*) is that we have *both* mixing *and* expansion.

Consider now processes II and II' in Fig. 4.3A. In process II we mix two components, A and B. For simplicity, we take $N = N_A = N_B$. In II' the two compartments contain only A particles. We remove the partition separating two compartments. In process II we find

$$\Delta S(II) = k_B N_A \ln 2 + k_B N_B \ln 2 = 2N k_B \ln 2 > 0$$
(4.41)

In II', on the other hand, the entropy change is zero.

$$\Delta S(II') = 0 \qquad (4.42)$$

The thermodynamic view is that in process II, we *see* mixing, on the other hand in process *II'*, nothing is observed to happen. At the molecular level the picture is quite different. In II, each of the components has expanded from V into $2V$. As discussed above, the positive entropy change in (4.41) is due to two *expansion* processes ($II_1 + II_2$). On the other hand, in II' two processes occur. First, the volume accessible to each particle has increased from V to $2V$, giving rise to an entropy change of $\Delta S(expansion) = 2N k_B \ln 2$. In addition, the assimilation term has also changed. Initially, each particle is assimilated by N particles, but in the final state by $2N$ particles. The resulting change in entropy is referred to as the assimilation entropy,

$\Delta S(assimilation) = -k_B \ln (2N)!/(N!)^2$ which for large N is

$$\Delta S(assimilation) = -2Nk_B \ln 2 \qquad (4.43)$$

Therefore, the combined change in entropy for process II' is

$$\Delta S(II') = \Delta S(expansion) + \Delta S(assimilation)$$
$$= 2Nk_B \ln 2 - 2Nk_B \ln 2 = 0 \qquad (4.44)$$

The net change in entropy is zero. Note, however that the cancellation in (4.42) is between two quantities that are of equal magnitude but of different origin.

4.4 Chemical Equilibrium in an Ideal-Gas Mixture

One of the most important aspects of the application of statistical thermodynamics is that it provides a method of calculating the equilibrium constant of a chemical reaction. From the general statistical thermodynamics expression for the chemical potential, we can obtain a general expression for the equilibrium constant. In some simple cases a numerical value of the equilibrium constant can be computed, and this value may be compared with the corresponding experimental value. In more complex systems such calculations are not feasible. Nevertheless, the general expressions are still useful for qualitative predictions of expected trends of changes in the equilibrium constant upon changing some parameters of the system.

Because of their central importance to almost all applications of statistical thermodynamics to chemistry and

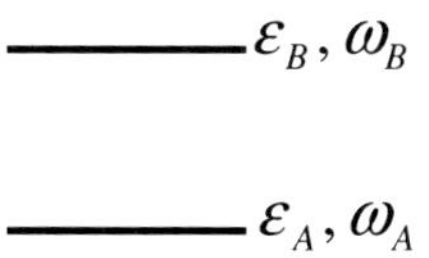

$$\underline{\qquad\qquad}\,\varepsilon_B,\omega_B$$

$$\underline{\qquad\qquad}\,\varepsilon_A,\omega_A$$

Fig. 4.4 Two energy levels ε_A and ε_B with corresponding degeneracies ω_A and ω_B.

biochemistry, we shall study in detail some aspects of systems in chemical equilibrium.

4.4.1 *Simple Isomerization Equilibrium*

We start with the simplest case of a chemical reaction. Consider a system of N particles in an ideal-gas phase of volume V and temperature T. We assume that each particle has only two internal (say, electronic or vibrational) energy levels, which we denote by ε_A and ε_B, with degeneracies ω_A and ω_B, respectively. No other internal degrees of freedom are ascribed to the particles (Fig. 4.4).

The partition function for such a system is

$$Q(T,V,N) = \frac{q^N V^N}{N!\Lambda^{3N}} \qquad (4.45)$$

where the internal PF is simply

$$q = \omega_A e^{-\beta\varepsilon_A} + \omega_B e^{-\beta\varepsilon_B} = q_A + q_B \qquad (4.46)$$

Here, q_A and q_B are the internal PFs of a single particle in state A and state B, respectively. Thus, if we know the molecular parameters: $m, \omega_A, \omega_B, \varepsilon_A, \varepsilon_B$ we can compute all the relevant thermodynamic quantities of our system.

Until this point we have considered a system of *one* component. No chemical equilibrium was mentioned. Now we choose to view the *same* system from a *different* point of view. Each molecule of the system can be in one of the two states i.e. A or B. We call a molecule in state A an A-molecule, and a molecule in state B a B-molecule. We stress that this classification of molecules into two groups is purely theoretical. It does not depend on whether or not we can detect or isolate the A molecules or the B molecules.

We denote by

$$Y_A = \frac{q_A V}{\Lambda^3}, \quad Y_B = \frac{q_B V}{\Lambda^3} \quad (4.47)$$

and

$$Y = Y_A + Y_B \quad (4.48)$$

With this notation we can rewrite the PF (4.45) as

$$Q(T, V, N) = \frac{Y^N}{N!} = \frac{(Y_A + Y_B)^N}{N!}$$

$$= \frac{1}{N!} \sum_{N_A=0}^{N} \frac{N!}{N_A! N_B!} Y_A^{N_A} Y_B^{N_B} \quad (4.49)$$

where we used the binomial expansion. The summation is over all possible values of N_A (where N_B is defined by $N_B = N - N_A$).

Next, we define the quantity

$$Q(T, V, N_A, N_B) = \frac{Y_A^{N_A} Y_B^{N_B}}{N_A! N_B!} \quad (4.50)$$

and rewrite (4.49) as

$$Q(T, V, N) = \sum_{N_A=0} Q(T, V, N_A, N_B) \quad (4.51)$$

We identify $Q(T, V, N_A, N_B)$ in (4.50) as the PF of a *two-component mixture* of A and B in an ideal gas. How did we get a two-component system from a one-component system? The partition function of our system is (4.51), not (4.50). The summation over N_A in (4.51) makes the sum (4.51) independent of N_A, so that the resulting sum on the rhs of (4.51) is a function of N only. However, there exists a particular value of N_A (and N_B) for which $Q(T, V, N)$ may be approximated by $Q(T, V, N_A, N_B)$. This is an important property of the PF of a macroscopic system at equilibrium. We shall examine this aspect shortly after we have introduced a few concepts.

According to the general rule of statistical thermodynamics, each term in the PF is proportional to the probability of the event characterized by that term. Here, $Q(T, V, N_A, N_B)$ is proportional to the probability of finding our (one-component) system in a state such that precisely N_A molecules are in state A and the rest, N_B molecules, in state B, i.e.

$$Pr(N_A) = \frac{Q(T, V, N_A, N_B)}{Q(T, V, N)} \qquad (4.52)$$

In this system we can calculate three specific values of N_A:

1. The most probable value of N_A, denoted N_A^*.
2. The average value of N_A, denoted $\langle N_A \rangle$.
3. The equilibrium value of N_A, denoted N_A^e.

These are the three conceptually different quantities. As we will soon see, their numerical values are, from the macroscopic point of view, identical.

The most probable value of N_A is defined as the value of N_A for which

$$\frac{\partial Pr(N_A)}{\partial N_A} = 0 \qquad (4.53)$$

It is easy to see that $Pr(N_A)$ has one extremum, which is a maximum. From (4.50) and (4.53) we obtain, using the Stirling approximation

$$\frac{\partial \ln Pr(N_A)}{\partial N_A} = \ln Y_A + \ln Y_B - \ln N_A$$

$$+ \ln (N - N_A) = 0 \qquad (4.54)$$

Solving (4.54) for N_A^*, we obtain

$$N_A^* = N \frac{Y_A}{Y_A + Y_B} \qquad (4.55)$$

For $N_A = N_A^*$, we also have

$$\frac{\partial^2 \ln Pr(N_A)}{\partial N_A^2} = - \left[\frac{1}{N_A} + \frac{1}{N - N_A} \right]_{N_A=N_A^*} < 0. \qquad (4.56)$$

which means that $\ln Pr(N_A)$, as well as $Pr(N_A)$, has a *maximum* at $N_A = N_A^*$.

The average value of N_A is defined by

$$\langle N_A \rangle = \sum_{N_A=0}^{N} N_A Pr(N_A)$$

$$= \frac{1}{Q(T, V, N)} \sum_{N_A} N_A \frac{Y_A^{N_A} Y_B^{N_B}}{N_A! N_B!} \qquad (4.57)$$

To evaluate the sum on the rhs of (4.57) we note the identity

$$Y_A \frac{\partial}{\partial Y_A} \left[\sum_{N_A} \frac{Y_A^{N_A} Y_B^{N_B}}{N_A! N_B!} \right] = \sum_{N_A} N_A \frac{Y_A^{N_A} Y_B^{N_B}}{N_A! N_B!} \qquad (4.58)$$

Hence from (4.57) and (4.58) we obtain

$$\langle N_A \rangle = \frac{Y_A}{Q} \frac{\partial}{\partial Y_A} \left[\sum_{N_A} N_A \frac{Y_A^{N_A} Y_B^{N_B}}{N_A! N_B!} \right]$$

$$= \frac{Y_A}{Q} \frac{\partial}{\partial Y_A} \left[\frac{(Y_A + Y_B)^N}{N!} \right]$$

$$= \frac{Y_A}{Q} \left[\frac{N(Y_A + Y_B)^{N-1}}{N!} \right] = N \frac{Y_A}{Y_A + Y_B} \qquad (4.59)$$

We find that the value of $\langle N_A \rangle$ in (4.59) is equal to the value of N_A^* in (4.55).

Exercise: The equilibrium value of N_A is defined by

$$\mu_A(N_A^e) = \mu_B(N_B^e) \qquad (4.60)$$

Use the expression of the chemical potential to solve Eq. (4.60) and show that N_A^e is equal to N_A^*, as well as to $\langle N_A \rangle$.

4.4.2 Thermodynamics of Isomerization Reaction

In this section we discuss the thermodynamics of a simple isomerization reaction

$$A \rightleftarrows B \qquad (4.61)$$

As before, we assume for simplicity that A and B have each one energy level; ε_A and ε_B respectively, and the corresponding degeneracies ω_A and ω_B, respectively. We are interested in the thermodynamics of this reaction.

The standard Helmholtz energy of this reaction is defined as the change in the Helmholtz energy for the conversion of one A molecule into one B molecule, keeping the temperature and the volume of the system fixed. We also assume that our system is an ideal gas.

The chemical potentials of A and B are given by

$$\mu_A = k_B T \ln \rho_A \Lambda_A^3 q_A^{-1} = \mu_A^* + k_B T \ln \rho_A \Lambda_A^3 \quad (4.62)$$

$$\mu_B = k_B T \ln \rho_B \Lambda_B^3 q_B^{-1} = \mu_B^* + k_B T \ln \rho_B \Lambda_B^3 \quad (4.63)$$

where ρ_α is the number density N_α / V, Λ_α^3 is the momentum partition function, q_α is the internal partition function of the species α, and μ_α^* is the pseudo-chemical potential of the species α.

Since A and B have the same mass, their momentum partition functions are equal, i.e.

$$\Lambda_A^3 = \Lambda_B^3 \quad (4.64)$$

At equilibrium we have the condition

$$\mu_A = \mu_B \quad (4.65)$$

From (4.62) to (4.65) we get the expression for the equilibrium constant of the reaction (4.61) in an ideal gas phase

$$K^{ig} = \left(\frac{\rho_B}{\rho_A} \right)_{eq} = \exp[-\beta(\mu_B^* - \mu_A^*)] \quad (4.66)$$

In this particular case the internal partition functions of the species A and B are

$$q_A = \omega_A \exp[-\beta \varepsilon_A], \quad q_B = \omega_B \exp[-\beta \varepsilon_B] \quad (4.67)$$

Hence

$$K^{ig} = \left(\frac{q_B}{q_B}\right) = \frac{\omega_B}{\omega_A} \exp[-\beta(\varepsilon_B - \varepsilon_A)] \quad (4.68)$$

Define the *standard Helmholtz energy* of the reaction by

$$\Delta A^\circ (A \to B) = \mu_B^* - \mu_A^* = -k_B T \ln K^{ig}$$

$$= (\varepsilon_B - \varepsilon_A) + k_B T \ln \frac{\omega_B}{\omega_A} \quad (4.69)$$

The corresponding *standard entropy* and *energy* (or enthalpy) changes are

$$\Delta S^\circ (A \to B) = -\frac{\partial \Delta A^\circ}{\partial T} = k_B \ln \frac{\omega_B}{\omega_A} \quad (4.70)$$

$$\Delta E^\circ (A \to B) = \Delta A^\circ + T \Delta S^\circ = \varepsilon_B - \varepsilon_A \quad (4.71)$$

The most important aspect of this simple reaction is that the energy change and the entropy change are completely independent of each other. The energy change is simply the difference in the energy levels, and the entropy change is related to the ratio of the degeneracies of the two energy levels. This neat separation between energy levels and degeneracies no longer holds when a solvent is present.

We can rewrite the equilibrium constant for this reaction as

$$K^{ig} = \exp[-\beta(\Delta E^\circ - T\Delta S^\circ)] \qquad (4.72)$$

Assuming that $\varepsilon_B - \varepsilon_A > 0$ and $\omega_B/\omega_A > 1$, we have $\Delta E^\circ > 0$ and $\Delta S^\circ > 0$. We can predict the following behavior of the equilibrium constants as we change the temperature. At very low temperatures $T \to 0$, the term $T\Delta S^\circ$ will be negligible. The equilibrium constant will depend only on the difference in the energy levels

$$K^{ig} = \exp[-\beta\Delta E^\circ] \to 0 \qquad (4.73)$$

Hence, the relative population of the B molecules will tend to zero. On the other hand, at very high temperatures $(T \to \infty)$, the term $T\Delta S^\circ$ will become larger than ΔE°. At this limit the equilibrium constant will be determined by the ratio of the degenaracies

$$K^{ig} \approx \exp\left[\frac{\Delta S^\circ}{k}\right] = \frac{\omega_B}{\omega_A} \qquad (4.74)$$

Figure 4.5 shows the dependence of the mole fraction $x_A = N_A^e/(N_A^e + N_B^e) = (1 + K)^{-1}$ on the temperature T for fixed value of $\varepsilon_B - \varepsilon_A$ and varying values of ΔS°.

As expected, all curves start with $x_A = 1$ at $T \to 0$. At $T \to \infty$, the limiting value depends on ω_B/ω_A. For $\omega_B/\omega_A = 1$, $x_A \to \frac{1}{2}$, and for $\omega_B/\omega_A = 1000$, $x_A \to (1000)^{-1}$ at this limit. These curves are very typical for any system at chemical equilibrium. Note that when ω_B/ω_B is very large the curve changes very sharply from $x_A = 1$ to $x_A = 0$.

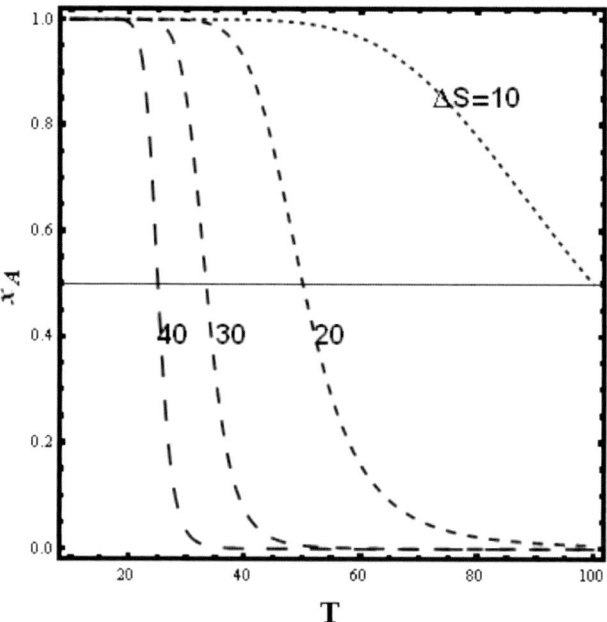

Fig. 4.5 The dependence of x_A on temperature T for a fixed value of $\Delta E = \varepsilon_B - \varepsilon_A = 1000$ and varying values of $\Delta S = k_B \ln \frac{\omega_B}{\omega_A}$.

4.4.3 *Generalizations*

We discuss here two generalizations of the simple example treated in the preceding section. First, consider again an isomerization reaction. This could be a *cis-trans* conversion in dichloroethylene or a *helix-coil* transition in a polypeptide. Symbolically we write

$$A \rightleftarrows B \qquad\qquad (4.75)$$

where now, instead of two internal energy levels, ε_A and ε_B as in (4.46), each molecule has a series of energy levels, depicted schematically in Fig. 4.6.

In some practical cases, the set of all states of a single molecule may be split into two groups in such a way that

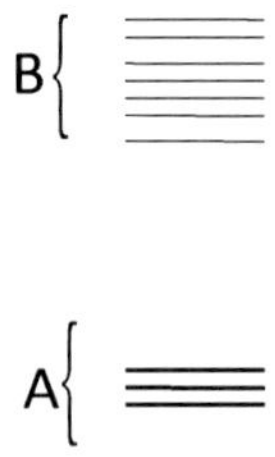

Fig. 4.6 Energy levels of a molecule which can be grouped into two groups A and B.

transitions between states in one group are much faster than transitions between states belonging to different groups. We denote by $\{i \in A\}$ all the internal states that belong to one group, A, and by $\{i \in B\}$ all the states belonging to group B. In this way we have chosen to view the one-component system of molecules as a mixture of two components A and B with the corresponding internal partition functions q_A and q_B, defined by

$$q = \sum_i \exp\left(-\beta\varepsilon_i\right) = \sum_{i \in A} \exp\left(-\beta\varepsilon_i\right)$$

$$+ \sum_{i \in B} \exp\left(-\beta\varepsilon_i\right) = q_A + q_B \qquad (4.76)$$

If the transitions between states belonging to different groups are extremely slow, then the system behaves like a mixture of two independent components. In principle we can even separate the system into two pure components or prepare a mixture with any composition of N_A and N_B. However, if the transitions are fast, then the system is essentially a one-component system and each molecule has an internal partition function q. From the theoretical point of view, any division of the set of the states into two groups defines a mixture-model. Such a division is independent of the rate of transitions between the states and of the distribution of the states. The choice of grouping is arbitrary. We define the corresponding canonical PF as

$$Q(T, V, N) = \frac{q^N V^N}{N! \Lambda^{3N}} = \frac{V^N}{N! \Lambda^{3N}} (q_A + q_B)^N$$

$$= \frac{V^N}{N! \Lambda^{3N}} \sum_{N_A} \frac{N!}{N_A! N_B!} q_A^{N_A} q_B^{N_B}$$

$$= \sum_{N_A} Q(T, V, N_A, N_B) \qquad (4.77)$$

where $Q(T, V, N_A, N_B)$ is the partition function of a mixture of N_A molecules of A and N_B molecules of B in the presence of an inhibitor which prevents the transition between the groups A and B. The sum over all possible N_A brings us back to the one-component system. Note again that the mixture-model point of view is independent of whether or not such an inhibitor actually exists.

We see that (4.77) is formally the same as (4.51) except for the new interpretation of q_A and q_B in (4.76). Therefore, we can proceed to obtain all the thermodynamic results in the same formal manner as in the simpler case, provided we keep in mind the new interpretation of q_A and q_B. Thus, the equilibrium values of N_A and N_B are

$$N_A^e = N \frac{Y_A}{Y_A + Y_B}, \quad N_B^e = N \frac{Y_B}{Y_A + Y_B} \qquad (4.78)$$

where

$$Y_A = \frac{q_A V}{\Lambda_A^3}, \quad Y_B = \frac{q_B V}{\Lambda_B^3} \tag{4.79}$$

The equilibrium constant is

$$K^{ig} = \frac{N_B^e}{N_A^e} = \frac{q_B}{q_A} \tag{4.80}$$

which is formally the same as (4.68).

The pseudo-chemical potentials for this case are

$$\mu_A^* = -k_B T \ln q_A = -k_B T \ln \sum_{i \in A} \exp[-\beta \varepsilon_i] \tag{4.81}$$

$$\mu_B^* = -k_B T \ln q_B = -k_B T \ln \sum_{i \in B} \exp[-\beta \varepsilon_i] \tag{4.82}$$

The pseudo partial molar entropy and energy (or enthalpy) are in this case

$$-S_A^* = \frac{\partial \mu_A^*}{\partial T} = -k_B \ln q_A - k_B T \frac{\partial \ln q_A}{\partial T} \tag{4.83}$$

$$E_A^* = \mu_A^* + T S_A^* \tag{4.84}$$

Denote by $p(i|A)$ the conditional probability of finding the state i given that the molecule is in one of the states in group A, i.e.

$$p(i|A) = \frac{\exp[-\beta \varepsilon_i]}{q_A} \tag{4.85}$$

With this definition we can rewrite the entropy S_A^* and the energy E_A^* in (4.83) and (4.84) as

$$S_A^* = k_B \ln q_A + \frac{\sum_{i \in A}[-\beta \varepsilon_i]}{q_A T}$$

$$= -k_B \sum_{i \in A} p(i|A) \ln p(i|A) \tag{4.86}$$

$$E_A^* = \sum_{i \in A} P(i|A) \varepsilon_i \tag{4.87}$$

Hence, the standard thermodynamic quantities for this case are

$$\Delta A^\circ(A \to B) = \mu_B^* - \mu_A^* = \Delta E^\circ(A \to B)$$
$$- T \Delta S^\circ(A \to B) \tag{4.88}$$

$$\Delta S^\circ(A \to B) = S_B^* - S_A^* = -k_B \sum_{i \in B} p(i|B) \ln p(i|B)$$

$$+ k_B \sum_{i \in A} p(i|A) \ln p(i|A) \tag{4.89}$$

$$\Delta E^\circ(A \to B) = E_B^* - E_A^* = \sum_{i \in B} p(i|B) \varepsilon_i$$

$$- \sum_{i \in A} p(i|A) \varepsilon_i = \langle \varepsilon_B \rangle - \langle \varepsilon_A \rangle \tag{4.90}$$

Compare these relationships with (4.69), (4.70) and (4.71). Clearly, when each group consists of only one energy level the last three relationships reduce to the corresponding relationships (4.69) to (4.71). Note also that the entropy difference is simply the difference in the Shannon measure of information (SMI) defined on the distributions $\{p(i|A)\}$ and $\{p(i|B)\}$, respectively.

In biochemistry, when dealing with huge molecules such as proteins or nucleic acids, it is often convenient to treat the system as being a quasi-two-state system. For instance, a protein may be viewed as having two states; the folded F (or native), and the unfolded U (or denatured) states. Clearly, each of these states consists of many quantum-mechanical states. It is convenient to define the Helmholtz energy of each of these combined states as follows:

$$A_U = -k_B T \ln q_U = -k_B T \ln \sum_{i \in U} \exp[-\beta \varepsilon_i] \quad (4.91)$$

$$A_F = -k_B T \ln q_F = -k_B T \ln \sum_{i \in F} \exp[-\beta \varepsilon_i] \quad (4.92)$$

These Helmholtz energies are then used as the "energy levels" of the protein. The equilibrium constant is thus

$$K^{ig} = \frac{q_U}{q_F} = \exp[-\beta(A_U - A_F)] \quad (4.93)$$

The standard Helmholtz energy of this reaction is simply

$$\Delta A^\circ = -k_B T \ln K^{ig} = A_U - A_F \quad (4.94)$$

One final comment about biomolecules is in order. In all the discussions above we assume that all the energy levels are accessible. In other words, there are transitions between all states. In large biomolecules there could be sets of energy levels which are separated, and such that there are no transitions between different sets of energy levels. In such case one must consider only those energy levels that are accessible at some given T, P, N.

The second generalization is a reaction of the type

$$v_A A + v_B B \rightleftarrows v_C C + v_D D \qquad (4.95)$$

where v_i is the stoichiometric coefficient of the ith component.

The PF of a system containing A, B, C and D at equilibrium is quite complicated. However, we already know that we can "freeze in" the equilibrium and treat the system as if it were a mixture of ideal gases. Thus, for any composition of N_A, N_B, N_C and N_D the PF of the "frozen-in" system is

$$Q(T, V, N_A, N_B, N_C, N_D) = \prod_i \frac{(q_i V)^{N_i}}{\Lambda_i^{3N_i} N_i!} \qquad (4.96)$$

where the product over i is over all species $i = A, B, C, D$.

In order to obtain the equilibrium constant, we use the thermodynamic condition of chemical equilibrium. First, we write the chemical potential of each species i as

$$\mu_i = k_B T \ln \rho_i \Lambda_i^3 q_i^{-1} = \mu_i^0 + k_B T \ln \rho_i \qquad (4.97)$$

where q_i is the internal PF of the molecule of species i. The energy levels for all species are measured relative to some common zero-energy level. The thermodynamic equilibrium condition is

$$v_A \mu_A + v_B \mu_B = v_C \mu_C + v_D \mu_D \qquad (4.98)$$

which leads to the equilibrium constant

$$K_\rho^{ig} = \exp\{-\beta[v_C \mu_C^0 + v_D \mu_D^0 - v_A \mu_A^0 - v_B \mu_B^0]\} \qquad (4.99)$$

or

$$K_\rho^{ig} = \prod_i \rho_i^{v_i} = \prod_i \left(\frac{q_i}{\Lambda_i^3}\right)^{v_i} \qquad (4.100)$$

Here, we have expressed the equilibrium constant in terms of the densities ρ_i. A more common expression for the equilibrium constant in an ideal-gas mixture is in terms of the partial pressures of the species. Thus, substituting $P_i = k_B T \rho_i$ in (4.100), we obtain

$$K_\rho^{ig} = \prod_i \rho_i^{v_i} = \prod_i (P_i/k_B T)^{v_i} = (k_B T)^{-\Sigma v_i} \prod_i P_i^{v_i}$$

$$= (k_B T)^{-\Sigma v_i} K_P^{ig} \qquad (4.101)$$

Note that in (4.100) and (4.101) we use the convention that v_i is positive for a product species (C and D) and negative for a reactant species (A and B).

4.4.4 Heat Capacity of a System in Chemical Equilibrium

The heat capacity of a system is an important thermodynamic quantity. In many cases one finds experimentally a very high value of the heat capacity, far larger than the value expected from the sum of the contributions of the heat capacity of each molecule in the system.

In this section we discuss a simple example which reveals the source of the high value of heat capacity. Again, we have N molecules in an ideal-gas phase, and each molecule has an internal PF given by

$$q = \sum_i \exp(-\beta \varepsilon_i) \tag{4.102}$$

The summation in (4.102) is over *all* states of a single molecule (excluding the momentum PF). The PF of the system is

$$Q(T, V, N) = \frac{q^N V^N}{N! \Lambda^{3N}} \tag{4.103}$$

The internal energy is

$$E = -T^2 \frac{\partial(A/T)}{\partial T} = \frac{3}{2} N k_B T + N \sum_i p_i \varepsilon_i \tag{4.104}$$

Here, each molecule has a kinetic energy of $\frac{3}{2} k_B T$, and an average internal energy $\sum p_i \varepsilon_i$, where the sum is over *all* internal states of a single molecule.

The constant-volume heat capacity is given by

$$C_V = \left(\frac{\partial E}{\partial T} \right)_{V,N} = \frac{3}{2} N k_B$$

$$+ \frac{N}{k_B T^2} \left[\sum_i p_i \varepsilon_i^2 - \left(\sum_i p_i \varepsilon_i \right)^2 \right]$$

$$= \frac{3}{2} N k_B + \frac{N}{k_B T^2} [\langle \varepsilon^2 \rangle - \langle \varepsilon \rangle^2] \tag{4.105}$$

This is the general expression for the heat capacity.

Now we choose to view our system as a two-component system. We follow the same grouping of all states into two groups as in Section 4.4.3 and rewrite (4.104) as

$$E = \frac{3}{2}Nk_BT + N\sum_i p_i\varepsilon_i = \frac{3}{2}Nk_BT$$

$$+N\left\{\sum_{i\in A}\frac{\varepsilon_i\exp[-\beta\varepsilon_i]}{q} + \sum_{i\in B}\frac{\varepsilon_i\exp[-\beta\varepsilon_i]}{q}\right\}$$

$$= \frac{3}{2}Nk_BT + \bar{N}_A\langle\varepsilon_A\rangle + \bar{N}_B\langle\varepsilon_B\rangle \qquad (4.106)$$

where $\langle\varepsilon_A\rangle$ and $\langle\varepsilon_B\rangle$ are the average energy levels of each group, as defined in (4.90), and $\bar{N}_A$ and $\bar{N}_B$ are the average number of molecules in each group defined by $\bar{N}_A = Nq_A$ and $\bar{N}_B = Nq_B$, respectively. The heat capacity in this representation is now

$$C_V = \frac{3}{2}Nk_B + \bar{N}_A\left(\frac{\partial\langle\varepsilon_A\rangle}{\partial T}\right) + \bar{N}_B\left(\frac{\partial\langle\varepsilon_B\rangle}{\partial T}\right)$$

$$+[\langle\varepsilon_B\rangle - \langle\varepsilon_A\rangle]\left(\frac{\partial\bar{N}_B}{\partial T}\right) \qquad (4.107)$$

The first term on the rhs of (4.107) is, as in (4.105), due to the translational degrees of freedom. The significance of the other two terms is as follows: Suppose that the transition between states belonging to two groups is forbidden, say by introducing an inhibitor that prevents such transitions; then $\bar{N}_A$ and $\bar{N}_B$ are fixed quantities. In this case, the last term on the rhs of (4.107) will not feature. If we allow transitions between the two groups (say by removing the inhibitor, or introducing a catalyst), then $\bar{N}_A$ and $\bar{N}_B$ are the equilibrium values of N_A and N_B, hence they are temperature dependent. The last term on the rhs of (4.107) is the change in the heat capacity of the system caused by introducing the catalyst.

We denote this contribution by

$$\Delta C^r = [\langle \varepsilon_B \rangle - \langle \varepsilon_A \rangle] \left(\frac{\partial \bar{N}_B}{\partial T} \right)$$

We now show that this term is always positive, i.e. introducing the catalyst to the system always *increases* the heat capacity of the system.

In the case of an ideal gas it is easy to calculate the temperature derivative of $\bar{N}_B$. For instance from

$$\bar{N}_B = \frac{q_B}{q} N \qquad (4.108)$$

we obtain

$$\frac{\partial \bar{N}_B}{\partial T} = \frac{N x_A x_B}{k_B T^2} [\langle \varepsilon_B \rangle - \langle \varepsilon_A \rangle] \qquad (4.109)$$

Hence

$$\Delta C^r = \frac{N x_A x_B}{k_B T^2} [\langle \varepsilon_B \rangle - \langle \varepsilon_A \rangle]^2 \geq 0 \qquad (4.110)$$

where x_A and x_B are the mole fractions of A and B at equilibrium.

The result (4.110) obtained here is for a special case of an ideal-gas mixture. We shall encounter a similar result for a more general case in Chapter 10.

4.5 Ideal Gas in a Gravitational Field

We consider an ideal gas (i.e. no interactions among the particles) consisting of N molecules at a temperature T contained in a vertical column of height H and cross section a (Fig. 4.7). The system is subjected to an external field of force (either gravitational or centrifugal) such that at each height z, the interaction between the particle of mass m and the external field is

$$U(z) = mgz \text{ (in a gravitational field)} \qquad (4.111)$$

$$U(z) = -\frac{m}{2}w^2z^2 \text{ (in a centrifugal field)} \qquad (4.112)$$

Here, g is the gravitational acceleration, z is the height of the particle measured relative to the bottom of the column in the case of the gravitational field or the distance from the axis of rotation in the case of a centrifugal field, and w is the angular velocity of rotation of the centrifuge. In general, g depends on the distance from the center of the earth. However, since H is very small compared to the radius of the earth, one can assume that g is almost constant within

Fig. 4.7 A column of ideal gas in a gravitational field.

the thermodynamic system. Also, we can neglect gravitational forces due to other particles in the system. In writing (4.111) we have set arbitrarily the zero of the potential energy at $z = 0$. Since in both cases (4.111) and (4.112) the external field operates on each particle separately, the treatment of the two cases is similar. We shall pursue only the case of gravitational field.

The canonical partition function for the system in constant gravitational field is

$$Q(T, V, N; g) = \frac{1}{N! \Lambda^{3N}} \int \exp\left[-\sum_{i=1}^{N} mgz_i\right] dR^N$$

$$(4.113)$$

Here we assume that the particles are spherical and structure-less (i.e. they have no internal degrees of freedom). Note that the configurational PF includes only interaction between each particle and the external field. Since the integrand is a product of functions each of which depends only on the coordinates of one particle, the entire integral may be performed to obtain

$$Q(T, V, N; g) = \frac{a^N}{N! \Lambda^{3N}} \left[\int_0^H \exp(-\beta mgz)dz\right]^N$$

$$= \frac{a^N}{N! \Lambda^{3N}} \left[\frac{1 - \exp(-\beta mgH)}{\beta mg}\right]^N$$

$$= \frac{V^N}{N! \Lambda^{3N}} q_g^N \qquad (4.114)$$

The integration over x_i and y_i produced the area a, and the volume of the entire system is $V = a \times H$. In (4.114) we have used the quantity q_g

$$q_g = \frac{1}{H} \int_0^H \exp\left(-\beta mgz\right)dz \qquad (4.115)$$

This quantity may be viewed as an "internal" PF of a single molecule.

The chemical potential of the gas in the system is defined by

$$\mu = \left(\frac{\partial A}{\partial N}\right)_{T,V,g} = k_B T \ln\left(\rho \Lambda^3 q_g^{-1}\right) \qquad (4.116)$$

where $\rho = N/V$ is the bulk density of the entire system. Note that when $g \to 0$, $q_g = 1$ and $\mu = k_B T \ln \rho \Lambda^3$, which is the chemical potential of an ideal gas in the absence of an external field. For $\beta mgH \ll 1$, the first-order correction to the chemical potential is

$$\mu = k_B T \ln \rho \Lambda^3 + \frac{1}{2} mgH \qquad (4.117)$$

We next calculate the density distribution along the z-axis. To obtain this distribution, we divide the length of the system into M strips of size Δz, in such a way that in each strip of volume $\Delta V = a\Delta z$, the intensive variables are almost constant. But Δz should be large enough so that macroscopic thermodynamics applies to the volume $a\Delta z$. With these assumptions we rewrite the integral in (4.114) as

$$\left[\int_0^H \exp\left(-\beta mgz\right)\right]^N$$

$$= \left[\sum_{l=1}^M \int_{z_l}^{z_l+\Delta_z} \exp\left(-\beta mgz\right)dz\right]^N$$

$$= \left(\sum_{l=1}^M q_l\right)^N \qquad (4.118)$$

where in each strip we take a representative point, hence

$$q_l = \int_{z_l}^{z_l+\Delta_z} \exp\left(-\beta mgz\right)dz \approx \exp\left(-\beta mgz_l\right)\Delta z$$

$$(4.119)$$

and where $z_1 = 0$ and $z_M + \Delta z = H$.

$$\left(\sum_{l=1}^M q_l\right)^N = \sum_{N_1,N_2,\ldots,N_M} \frac{N!}{\prod N_l!} \prod_{l=1}^M q_l^{N_l} \qquad (4.120)$$

Hence, the PF (4.114) may be rewritten as

$$Q(T,V,N;g) = \sum_{\substack{N_1,N_2,\ldots,N_M \\ \Sigma N_l = N}} \prod_{l=1}^M \frac{(aq_l)^{N_l}}{N_l!\Lambda^{3N_l}} \qquad (4.121)$$

This has the form of an equilibrium distribution of the N particles into M species, the species being distinguished by their different heights z_i in the system. Each term corresponds to a specific set of $N_1 \cdots N_M$ where N_l is the number of particles in the lth strip and $\Sigma N_l = N$ the total number of particles in the system.

Let $\bar{N}_l$ be the average, or the equilibrium, number of particles in the lth strip. We now treat the lth strip as a macroscopic system, for which the PF is

$$Q_l(T, a\Delta z, \bar{N}_l; g) = \frac{(aq_l)^{\bar{N}_l}}{\bar{N}_l! \Lambda^{3\bar{N}_l}} \qquad (4.122)$$

The chemical potential of the gas at the l strip is defined by

$$\mu_l = k_B T \ln \frac{\bar{N}_l \Lambda^3}{aq_l} = k_B T \ln \rho_l \Lambda^3 + mgz_l \qquad (4.123)$$

In (4.123) ρ_l is the equilibrium density at height z_l or in the lth strip. Since at equilibrium the chemical potential is constant throughout the entire system, we must have the equalities:

$$\mu_1 = \mu_2 = \mu_3 = \cdots = \mu_M \qquad (4.124)$$

From this we can obtain the density distribution, i.e. from Eqs. (4.123) and (4.124) we obtain

$$\left(\frac{\rho_l}{\rho_1}\right)_{eq} = \exp[-\beta mg(z_l - z_1)] \qquad (4.125)$$

This is the well-known Boltzmann distribution of densities. The higher the strip, the less dense the gas is. Assuming that the ideal-gas law holds for each of the strips, we can write, for each l, the corresponding pressure

$$P_l = \rho_l k_B T \tag{4.126}$$

Hence

$$P_l = P_1 \exp(-\beta mg z_l) \tag{4.127}$$

which gives the pressure as a function of the height z_l, where P_1 is the pressure at the bottom of the system (i.e. at $z = 0$)

The probability of finding a particle at height between z and $z + dz$ is proportional to the density of particles at that height, i.e.

$$Pr(z)dz = \alpha \exp(-\beta mg z)dz \tag{4.128}$$

where α is the normalization constant, which is obtained from the condition

$$\int_0^H Pr(z)dz = 1 \tag{4.129}$$

Hence, we can solve for α

$$\alpha = \frac{\beta mg}{1 - \exp(-\beta mgH)} \tag{4.130}$$

The bulk density in the system is related to the densities ρ_l at each level by

$$\rho = \frac{N}{V} = \frac{\sum_{l=1}^{M} \bar{N}_l}{V} = \frac{a \Delta z}{V} \sum_l \rho_1 \exp(-\beta mg z_l) \tag{4.131}$$

From (4.114), (4.115) and (4.131) we obtain

$$\rho q_g^{-1} = \rho_1 \tag{4.132}$$

Substituting (4.132) in (4.116) we obtain

$$\mu = k_B T \ln \rho \Lambda^3 q_g^{-1} = k_B T \ln \rho_1 \Lambda^3 = \mu_1 \qquad (4.133)$$

where μ is the chemical potential of the bulk gas and μ_1 is the chemical potential of the gas in first ($l = 1$) strip. Because of Eq. (4.124), μ is also equal to any μ_l.

Next, we calculate some of the thermodynamic quantities of this system.

The Helmholtz energy of the entire system is given by

$$A = -k_B T \ln Q = -k_B T \ln \frac{V^N q_g^N}{N! \Lambda^{3N}}$$

$$= A(g = 0) - k_B T N \ln q_g \qquad (4.134)$$

where $A(g = 0)$ is the Helmholtz energy of the system in the absence of the gravitational field, i.e. $g = 0$.

The entropy of the system is

$$S = -\frac{\partial A}{\partial T} = S(g = 0) + k_B N \ln q_g + \frac{Nmg\bar{z}}{T}$$

$$\qquad (4.135)$$

where $\bar{z}$ is defined by

$$\bar{z} = \int_0^H Pr(z)z dz \qquad (4.136)$$

The energy of the system is

$$E = A + TS = E(g = 0) + Nmg\bar{z} = \frac{3}{2} Nk_B T + Nmg\bar{z}$$

$$\qquad (4.137)$$

The first term on the rhs of (4.137) is the average kinetic energy of the particles. The second term is the average potential energy of the particles in the gravitational field.

The heat capacity of the system is

$$C_V = \left(\frac{\partial E}{\partial T} \right)_{N,V} = \frac{3}{2} N k_B + N m g \frac{\partial \bar{z}}{\partial T}$$

$$= \frac{3}{2} N k_B + \frac{N}{k_B T^2} (mg)^2 \overline{(z - \bar{z})^2} = C_V(g = 0) + \Delta C_V$$

$$(4.138)$$

Since $\overline{(z - \bar{z})^2}$ is always positive, the heat capacity always increases upon turning on the gravitational field, i.e. $\Delta C_V > 0$. This result is similar to the one we encountered in connection with the heat capacity of a system in chemical equilibrium (Section 4.4.4).

Exercise: Calculate the change in entropy in the two processes of expansion of an ideal gas, described in Fig. 4.8. In the

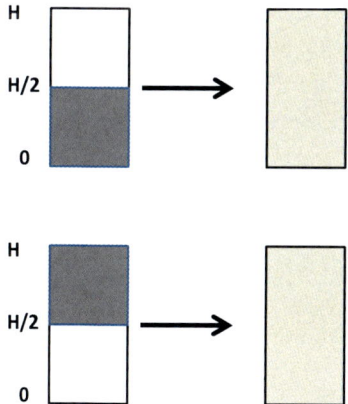

Fig. 4.8   Expansion of ideal gas in a gravitational field.

first process the gas expands "upwards," i.e. from the range of heights $0 \leq z \leq H/2$ to $0 \leq z \leq H$. In the second, the gas expands "downwards" from $H/2 \leq z \leq H$ to $0 \leq z \leq H$. The column of the gas is isolated, i.e. no exchange of heat with its surroundings. How will the temperature change in these two processes?

5

Systems of Adsorbed Non-Interacting Ligands

Binding of small molecules (ligands) to larger molecules abound in biochemistry. These phenomena range from binding of oxygen to hemoglobin to binding of drugs to proteins or to DNA, to binding of proteins to DNA in controlling the process of expression of specific genes.

In this chapter we will study the simplest case of binding systems. The typical binding curve, referred to as adsorption isotherm, or Langmuir isotherm is shown in Fig. 5.1.

The experimental observation is very simple. We have an adsorbing system, which could be a piece of a solid or a solution of polymer molecules. The surface of this material can adsorb ligands, which are at equilibrium with either a gaseous phase, or a solution at a given temperature T and pressure P. If we measure the quantity of adsorbed ligand on the surface, say by weighing the solid, we find that the weight, or the total quantity of the ligand adsorbed on the solid has

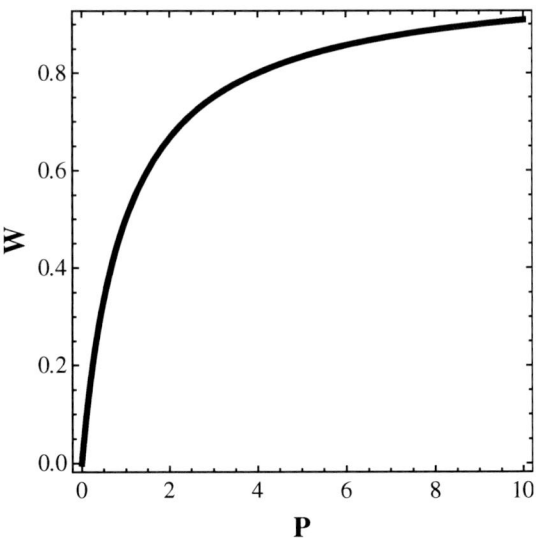

Fig. 5.1 The experimental dependence of the total weight of adsorbed gas (W) on the pressure (P).

the following functional form:

$$W = C \frac{KP}{1 + KP} \tag{5.1}$$

where W is the measured weight of the adsorbed ligand, and P is the partial pressure of the gas at equilibrium with the adsorbed molecules. C and K are constants, independent of P, but could depend on the temperature.

Our goal is to find a molecular model for the adsorption process that reproduces the same functional dependence of w on P as in (5.1), and thereby to provide a molecular interpretation of the quantities C and K in this equation. In Fig. 5.1, we plot $\theta = W/C$ as a function of the partial pressure of the ligand P.

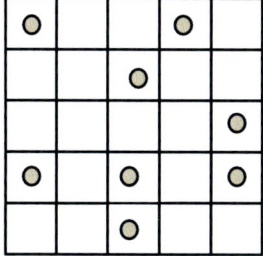

Fig. 5.2   Distribution of N ligands on M sites ($N = 8, M = 25$).

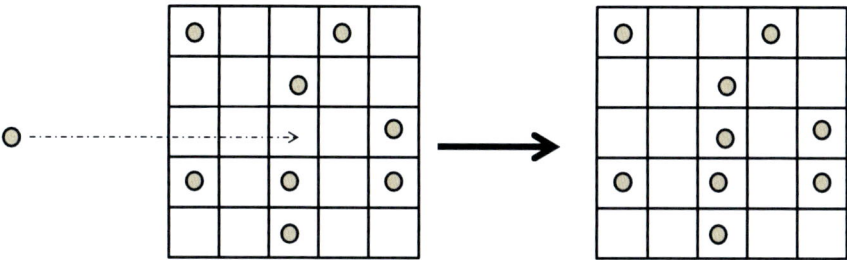

Fig. 5.3   The process of adding one ligand to a fixed site.

5.1 The Molecular Model and its Solution

Consider an adsorbing crystal, the surface of which provides M sites for adsorbing ligands of type A, Fig. 5.2. We make the following simplified assumptions: Each site can adsorb only one ligand A. The interaction between the site and the adsorbed molecule is characterized by a single parameter, denoted by U. This is defined as the energy change for bringing A from a fixed position at infinite separation from the crystal to a specific site on the surface of the crystal, Fig. 5.3. The sites are identical, but distinguishable; i.e. different configurations of the molecules on the M sites are distinguished as different states. The adsorbed molecules are indistinguishable, i.e. exchanging any two A molecules on two sites gives the same state of the system. The sites are independent. In general this assumption requires that the adsorption of A on one site does not affect the probability of adsorption on any other site. The adsorbed molecules A are in equilibrium with an ideal gas phase containing A molecules. We also assume that the internal degrees of freedom of either the ligands or the adsorbing crystal are not affected by the adsorption process.

With the above description of the model we turn to construct the partition function of our system. The system consists of a crystal having M sites and N ligands distributed on these sites ($N \leq M$). Since the crystal is presumed to be unaffected by the adsorption process, we may write the PF of the entire system (crystal and ligands) as

$$Q_{system} = Q_{cryst} Q_{ad} \qquad (5.2)$$

where Q_{cryst} is the PF of the empty crystal and Q_{ad} includes both the properties of the ligands and the interaction energy between the ligands and the crystal.

In this model, Q_{cryst} is unchanged in the process of adsorption. Therefore, we can ignore the factor Q_{cryst}. This is the same as treating the ratio Q_{system}/Q_{cryst} or the excess Helmholtz energy $A_{system} - A_{cryst}$. We will therefore write the PF of the adsorbed particles as

$$Q_{ad} = q_A^N \sum_E g(E) \exp\left(-\beta E\right) \qquad (5.3)$$

where q_A is the internal PF of a single A molecule and the sum is over all energy levels associated with different configurations of the A molecules on the surface of the crystal. In this model q_A does not change in the process of adsorption. Therefore, we factored the quantity q_A^N from the sum in (5.3).

In this particular model, there is only one energy level. Once we have fixed the number of adsorbed molecules N, the energy level is NU, the total interaction energy between all the ligands and all the sites. The only quantity that we need to compute is the configurational degeneracy $g(E)$, i.e. the number of distinguishable states that correspond to the single energy level $E = NU$. This number is simply the number of ways that N indistinguishable molecules can be distributed among M *distinguishable* sites, i.e.

$$g(E) = \frac{M!}{N!(M - N)} = \binom{M}{N} \qquad (5.4)$$

The counting of the number of configuration is the same as we have done in Section 3.5.

The PF of the adsorbed molecules is thus

$$Q_{ad}(T, M, N) = \frac{q_A^N M!}{N!(M - N)!} \exp[-\beta NU] \qquad (5.5)$$

This is an explicit function of the macroscopic parameters T, M, N, and the molecular parameters U, as well as those included in q_a.

5.2 Thermodynamics of the Adsorbed System and the Langmuir Isotherm

The Helmholtz energy for the adsorbed ligands is obtained from Eq. (5.5).

$$A_{ad} = -k_B T \ln Q_{ad}$$

$$= -k_B T \ln q_a + NU - k_B TM \ln M + k_B TN \ln N$$
$$+ k_B T (M - N) \ln (M - N) \qquad (5.6)$$

where we used the Stirling approximation for the factorial $[\ln N! \approx N \ln N - N]$.

The Helmholtz energy per site is

$$a_{ad} = \frac{A_{ad}}{M} = -k_B T \theta \ln q_A + \theta U$$
$$+ k_B T [\theta \ln \theta + (1 - \theta) \ln (1 - \theta)] \qquad (5.7)$$

where $\theta = N/M$ is the fraction of occupied sites, $(1 - \theta)$ being the fraction of empty sites

The entropy per site is

$$S_{ad} = \frac{S_{ad}}{M} = \frac{-1}{M} \left(\frac{\partial A_{ad}}{\partial T} \right)_{N,M}$$
$$= \frac{\partial}{\partial T} [k_B T \theta \ln q_A] - k_B [\theta \ln \theta + (1 - \theta) \ln (1 - \theta)]$$
$$(5.8)$$

For structureless ligands, i.e. for ligands having no internal degrees of freedom, $q_a = 1$ and (5.8) reduces to

$$S_{ad} = -k_B [\theta \ln \theta + (1 - \theta) \ln (1 - \theta)] \qquad (5.9)$$

Note that $\theta = N/M$ is the probability of finding a specific site occupied, and $(1 - \theta)$ is the probability of finding that site empty. Therefore, apart from the constant k_B and the choice of the base of the logarithm, the entropy of the adsorbed system is identical to the SMI of the system (see Section 3.5). This interpretation of the entropy as a particular case of SMI is discussed in Section 3.5. It should be noted that in some textbooks the quantity in (5.9) is interpreted as an "entropy of mixing". Clearly, this interpretation is faulty. First, because there is no process of mixing here; second, even if we conceive the system as a mixture of occupied and empty sites, the mixing of these two types of sites does not have any effect on the entropy of the system.

The chemical potential of the adsorbed molecules is

$$\mu_A^{ad} = \left(\frac{\partial A_{ad}}{\partial N}\right)_{T,M}$$

$$= -k_B T \ln q_A + U + k_B T \ln\left[\frac{\theta}{1-\theta}\right] \qquad (5.10)$$

We now examine two limiting cases: For $\theta \ll 1$, we have

$$\mu_A^{ad} = \mu_A^{0,ad} + k_B T \ln \theta \qquad (5.11)$$

This has the general form of a chemical potential in an ideal dilute solution, $\mu_A^{ad} \to \infty$, when $\theta \to 0$. The other extreme case is when $\theta \to 1$; then $\mu_A^{ad} + \infty$. The form of the function $\ln[\theta/(1-\theta)]$ is depicted in Fig. 5.4. At the limit $\theta \to 0$ there is an infinite "driving force" for A to flow from the gas phase to the sites. On the other hand, at the limit $\theta \to 1$, the sites are all occupied and there is an infinite repulsive force or resistance for accepting further molecules on the surface.

Next, we derive the adsorption isotherm. We assume that the ligand molecules are at equilibrium with an ideal gas phase. The chemical potential of the molecules in the gaseous phase is

$$\mu_A^g = k_B T \ln \rho_A \Lambda_A^3 q_A^{-1} \qquad (5.12)$$

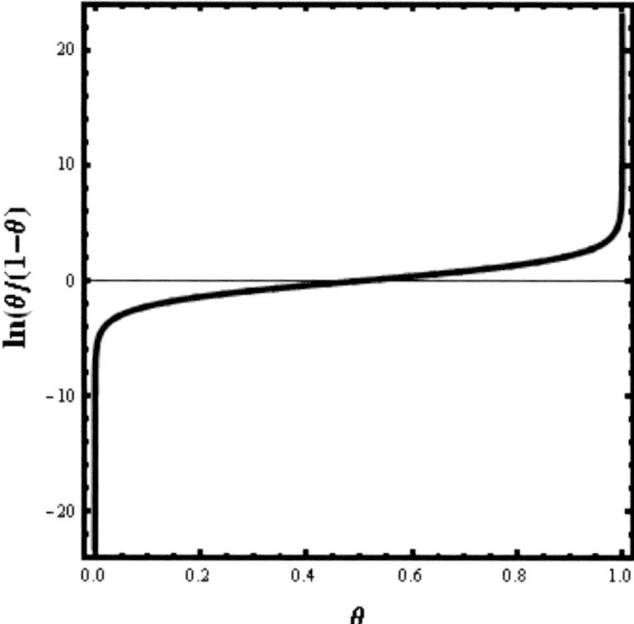

Fig. 5.4 The "driving force" for the adsorption of ligand as a function of θ.

According to the assumptions of the model, q_A in (5.12) is the same as q_A in (5.10), i.e. the internal degrees of freedom of A do not change in the process of adsorption. At equilibrium we have the equality.

$$\mu_A^{ad} = k_B T \ln \left(q_A^{-1} \frac{\theta}{1-\theta} \right) + U = \mu_A^g = k_B T \ln \rho_A \Lambda_A^3 q_A^{-1}$$

$$(5.13)$$

Hence, we can solve for θ to obtain

$$\theta = \frac{K' \rho_A}{1 + K' \rho_A}, \quad K' = \Lambda_A^3 \exp(-\beta U) \qquad (5.14)$$

This is essentially the Langmuir adsorption isotherm. We may rewrite it in terms of pressure P, instead of density ρ_A

by using the ideal-gas equation of state $P = k_B T \rho_A$, i.e.

$$N = M \frac{KP}{1 + KP}, \quad \text{where } K = \frac{K'}{k_B T} \quad (5.15)$$

Compare this result with the empirical function (5.1). What we have obtained is a molecular model that reproduces the same functional dependence of N on P. Here, M is identified with the total number of sites. In addition, we have an explicit expression for the constant K in terms of the molecular parameters of the system; these are the mass of each of the molecules, contained in Λ_A^3, and the adsorption energy U.

5.3 (***) Application of the Grand Partition Function for the Adsorbed System

In Section 5.2 we derived the Langmuir isotherm using the canonical PF. In this section we derive the same binding isotherm for the same system. There are two reasons for doing so. One, is to practice the application of the Grand PF and to show that one can obtain the same result from any PF. Second, as we will see in Chapter 6 that when there are interactions among the ligand, the application of the Grand PF becomes indispensable.

The Grand PF for our system is given by

$$\Xi(T, M, \mu) = \sum_{N=0}^{M} \lambda_A^N Q(T, M, N) \quad (5.16)$$

where $\lambda_A = \exp(\beta \mu)$ is the absolute activity of A: The sum over N means that we open the system with respect to A. This can be done, for instance, by letting the gaseous phase be at equilibrium with a reservoir of A molecules at a constant chemical potential μ. Each term in the PF is proportional to the probability of finding precisely N molecules on the adsorbing surface.

Using (5.5) in (5.16) we can sum over N to obtain the result

$$\Xi(T, M, \mu) = \sum_{N=0}^{M} \lambda_A^N q_A^N \frac{M!}{N!(M - N)!} \exp(-\beta N U)$$

$$= (1 + Y_A \lambda_A)^M = \xi^M \tag{5.17}$$

where

$$Y_A = q_A \exp(-\beta N U) \tag{5.18}$$

and

$$\xi = 1 + Y_A \lambda_A \tag{5.19}$$

ξ is referred to as the Grand PF of *one* site. Note that ξ has the same formal structure as the grand PF of the entire system, namely, it is a sum over all possible states of a *single* site; empty and occupied. The most important property of Ξ is its factorization into M equal factors (Eq. (5.17)). This feature follows from the assumption of the independence of the sites. The result (5.17) is important, and we will use it repeatedly in the following sections.

The average number of adsorbed molecules in this system is obtained from (see Section 1.4).

$$\bar{N} = k_B T \frac{\partial \ln \Xi}{\partial \mu} = \lambda_A \frac{\partial \ln \Xi}{\partial \lambda_A} = \lambda_A M \frac{\partial \ln \xi}{\partial \lambda_A} \tag{5.20}$$

or equivalently

$$\theta = \frac{\bar{N}}{M} = \frac{Y_A\lambda_A}{1 + Y_A\lambda_A} \qquad (5.21)$$

This is essentially the Langmuir isotherm. However, in contrast to Eqs. (5.14) or (5.15), this relationship is more general since it applies to any state of the A molecules at a given chemical potential μ (or equivalently, λ_A). No assumption of ideality of the gas at equilibrium with the surface is used.

If the ligands are at equilibrium with an ideal gas phase, then we have

$$\mu_A^g = k_B T \ln \rho_A \Lambda_A^3 q_A^{-1} \qquad (5.22)$$

or equivalently

$$\lambda_A = \rho_A \Lambda_A^3 q_A^{-1} \qquad (5.23)$$

Hence,

$$Y_A\lambda_A = \rho_A \Lambda_A^3 \exp(-\beta U) = K'\rho_A \qquad (5.24)$$

In this case (5.21) reduces to (5.14). Note than in (5.21) we have the *average* $\bar{N}$, whereas in (5.14) we had a *fixed* value of N. It is also useful to express the average quantity $\bar{N}$ in terms of the probability distribution. It follows from the general expression for $\bar{N}$ (Section 1.4) that

$$\bar{N} = k_B T \frac{\partial \ln \Xi}{\partial \mu} = \sum_{N=0}^{M} N Pr(N) \qquad (5.25)$$

where $Pr(N)$ is the probability of finding N particles on the surface. Likewise, $\bar{N}/M$ is interpreted as probability of finding a specific site occupied. From (5.21) we also have the following useful relationship

$$\theta = \frac{\bar{N}}{M} = \frac{Y_A \lambda_A}{\xi} = 0\frac{1}{\xi} + 1\frac{Y_A \lambda_A}{\xi}$$

$$= 0 \times Pr(0) + 1 \times Pr(1) \qquad (5.26)$$

Thus, the average occupation number of a single site is written in the same form as (5.25) but for a single site instead of the whole system. The probabilities of finding a specific site empty or occupied are read from the grand PF of a single site (5.19), i.e.

$$Pr(0) = \frac{1}{\xi} \qquad (5.27)$$

$$Pr(1) = Y_A \frac{\lambda_A}{\xi} = \frac{\bar{N}}{M} \qquad (5.28)$$

5.4 Probability and Thermodynamics of "Cavity" Formation

In this section we further elaborate on the definitions of the probabilities $Pr(0)$ and $Pr(1)$ and their relationship with the thermodynamics of adsorption. Although the relationship $Pr(1) = \bar{N}/M = \theta$ is intuitively clear, we derive it by using a different argument. In this section we return to the canonical ensemble, i.e. N is fixed. The quantity $\theta = N/M$ is the fraction of the occupied sites (in the open system the corresponding quantity $\bar{N}/M$ is the average fraction of the occupied sites). In this system, the probability of finding *any* site occupied is unity. If $N \neq 0$, there is always one site occupied. However, if we select a *specific site*, say the ith site, we can ask what the probability is of finding this specific site occupied. In our particular model the answer can be given using the so-called classical definition of probability; the probability Pr (ith site occupied) is equal to the total number of configurations that are consistent with the event "ith site occupied" divided by the total number of configurations of the system, i.e.

$$Pr(ith\ site\ occupied) = \frac{\binom{M-1}{N-1}}{\binom{M}{N}} = \frac{N}{M} = \theta \qquad (5.29)$$

The total number of configurations consistent with the requirement that the ith site be occupied is calculated by placing one ligand on the ith site and then counting all the possible configurations of the remaining $N - 1$ particles on the $M - 1$ sites.

The implicit assumptions made in calculating the probability in (5.29) is that each of the possible outcomes is an elementary event and that all elementary events have equal probabilities, i.e. each specific configuration of the system has the probability of $\binom{M}{N}^{-1}$.

This method of calculating probabilities is invalid when there are interactions among the ligands, in which case different configurations do not have the same probability. Nevertheless, the result (5.29) is still valid even when there are interactions. To prove this, we need a different argument which is presented below.

Suppose that we have labeled all the N ligands. Let Pr(kth ligand on the ith site) be the probability of finding a specific ligand say, the kth, on a specific site, say the ith site) Since the kth ligand must be found somewhere on one of the M sites, we must have the normalization condition

$$\sum_{i=1}^{M} Pr(k\text{th ligand on the }i\text{th site}) = 1 \qquad (5.30)$$

Since the M sites are equivalent the probabilities in (5.30) are independent of the index i. Therefore, the sum over i in (5.30) produces M equal terms.

$$\sum_{i=1}^{M} Pr(k\text{th ligand on the }i\text{th site})$$

$$= MPr(k\text{th ligand on the }i\text{th site}) = 1 \qquad (5.31)$$

From (5.31) it follows that

$$Pr(k\text{th ligand on the }i\text{th site}) = \frac{1}{M} \qquad (5.32)$$

Note that although this quantity is independent of the specific site, we still leave the index i, to stress that we are considering a *specific* site, the ith site.

Since the ligands are also equivalent the probabilities $Pr(k$th ligand on the ith site) are also independent of the index k. In addition, the N events "particle k occupies a specific site" are disjoint events ($k = 1, \ldots, N$); therefore, the probability of finding *any* of the N ligands on the specific site is the sum over k, i.e.

$$Pr(1) = Pr(any \ ligand \ on \ i)$$

$$= \sum_{k=1}^{M} Pr(k\text{th ligand on the }i\text{th site})$$

$$= \frac{N}{M} \tag{5.33}$$

This is the same result as (5.29). We have returned to the notation $Pr(1)$, eliminating the index i. It should be remembered however, that we always refer to a *specific* site. To conclude, the probability of finding *any* site empty or occupied is *unity* (if $0 < N < M$). The probabilities of finding a specific site empty or occupied are $(1 - \theta)$ and θ, respectively. These quantities are also related to the free energy change for the following process. Consider again the same system of N ligands on M sites. The Helmholtz energy of this system is (5.6)

$$A_{ad} = -k_B T \ln Q_{ad} \tag{5.34}$$

We now form "a cavity at i". By this we mean that we constrain the system in such a way that the ith site must always remain empty. We denote $A_{ad}(cav)$ and $Q_{ad}(cav)$ the Helmholtz energy and the PF of a system of N ligands on M sites with the stipulation that the ith site is empty. The corresponding relationship holds

$$A_{ad}(cav) = -k_B T \ln Q_{ad}(cav)$$

$$= -k_B T \ln \left[\frac{q_A^N (M-1)!}{N!(M-N-1)!} \exp(-\beta NU) \right] \tag{5.35}$$

Note that the energy U, and the number of ligands N in (5.35) are the same as in (5.5). The only difference is that the N ligands are now distributed over $M - 1$ sites. Therefore, the change in the Helmholtz energy for forming a "cavity" *at a specific site* is

$$\Delta A(cav \ at \ i) = A_{ad}(cav) - A_{ad}$$
$$= -k_B T \ln \frac{(M - 1)!(M - N)!}{(M - N - 1)!M!}$$
$$= -k_B T \ln (1 - \theta) \qquad (5.36)$$

This is an important result. The Helmholtz energy change for creating a "cavity at i" is related to the probability of finding a specific site empty. We note that in this particular model the Helmholtz energy change is purely an entropy effect, i.e.

$$\Delta S(cav \ at \ i) = -\frac{\partial \Delta A(cav \ at \ i)}{\partial T} = k_B \ln (1 - \theta) \qquad (5.37)$$

and

$$\Delta E(cav \ at \ i) = 0 \qquad (5.38)$$

There are always fewer possibilities of distributing N particles on $M - 1$ sites than on M sites. This is the reason for the negative entropy change (5.37). The corresponding energy change is obviously zero.

When interactions among the ligands exist, the method of calculation of $\Delta S(cav \ at \ i)$ as in (5.36) does not apply. In this case we do not have an explicit expression for either Q_{ad} or $Q_{ad}(cav)$. Nevertheless, the result $\Delta A(cav \ at \ i) = -k_B T \ln (1-\theta)$ obtained in (5.36) is still valid. The argument leading to this result when interaction exists is different from the one given above. In the most general case, the PFs of two systems with and without a cavity are defined by

$$Q_{ad}(cav) = \sum_{cav} \exp(-\beta E_j) \tag{5.39}$$

$$Q_{ad} = \sum_{all\ J} \exp(-\beta E_j) \tag{5.40}$$

In (5.39) the sum is over all states that are consistent with the presence of a cavity in the system. In (5.40) the sum is over all possible states of the system. Since the probability of finding any particular state j is $\exp(-\beta E_j)/Q_{ad}$, the probability of finding a "cavity" is the sum over the probabilities of all the (disjoint) events consistent with the requirement that such a cavity exists. i.e.

$$Pr(cav) = \sum_{cav} P_j = \frac{Q_{ad}(cav)}{Q_{ad}} \tag{5.41}$$

This is a very general result valid for a system with or without interactions, and also for any size of a cavity. In particular, if the size of the cavity is equal to one site only, say the ith site, then we already have Eq. (5.33) from which we infer that the probability of finding the site i empty is

$$Pr(0) = Pr(cav\ at\ i) = 1 - \frac{N}{M} = 1 - \theta \tag{5.42}$$

Combining (5.41) and (5.42) we obtain

$$\Delta A(cav\ at\ i) = -k_B T \ln(1 - \theta) \tag{5.43}$$

which is the same as (5.36). We stress again that this relationship is valid for a cavity having the size of one site only. If there are no interactions, it is easy to calculate the work required to create a cavity of any size. For instance, a cavity of two sites is discussed below. The situation is far more complicated in the case of interacting ligands.

Returning to our model let i and j be two specific sites. These could either be adjacent or far apart. The probability of finding these two sites occupied is denoted by $Pr(1, 1)$. This can easily be computed as follows

$$Pr(1, 1) = \frac{\binom{M-2}{N-2}}{\binom{M}{N}} = \frac{N(N-1)}{M(M-1)}$$

$$= \theta \frac{(N-1)}{(M-1)} \tag{5.44}$$

Equation (5.44) is computed by placing two particles at the specific sites, and distribute the remaining $N - 2$ particles over the remaining $M - 2$ sites.

Similarly, the probability of finding the two sites empty, i.e. a cavity of size two is

$$Pr(0, 0) = \frac{\binom{M-2}{N}}{\binom{M}{N}} = \frac{(M-N)(M-N-1)}{M(M-1)}$$

$$= (1 - \theta)\left(1 - \frac{N}{M-1}\right) \tag{5.45}$$

Here, where we distribute the N particles over the $M - 2$ available sites (i.e. excluding the ith and the jth sites that are left empty).

Recall that θ and $(1 - \theta)$ are the probabilities of finding a specific site occupied and empty, respectively. From (5.44) we see that even when there are no interactions, the probability $Pr(1, 1)$ is not exactly equal to the product of the probabilities of finding each of the sites occupied. This means the events: "site i occupied" and "site j occupied" are not independent events.

In the limit of the macroscopic system i.e. when $M \to \infty$ and $N \to \infty$, but N/M is constant, we obtain the complete independence of the two sites, i.e.

$$Pr(1, 1) = \theta^2 = (Pr(1))^2 \qquad (5.46)$$
$$Pr(0, 0) = (1 - \theta)^2 = (Pr(0))^2 \qquad (5.47)$$
$$Pr(1, 0) = \theta(1 - \theta) = Pr(0)Pr(1) \qquad (5.48)$$

The difference between, say (5.44) and (5.46) arises from the finite size of the system.

In the next chapter we will discuss pair distributions of the type $Pr(1, 1)$, $Pr(0, 0)$ or $Pr(1, 0)$ for which factorization of the type (5.46) to (5.48) is not valid. These pair distributions will be important in the study of cooperativity phenomena in binding systems.

5.5 The Analogue of the Solvation Process in Binding Systems

In Section 4.2 we introduced the concept of the pseudo-chemical potential of an ideal gas. A similar concept may be defined for the adsorbed molecules. The pseudo-chemical potential is defined as the change in the Helmholtz energy for the energy for the process of placing a ligand at a *fixed* site, say the jth.

$$\mu_A^*(at\ j) = -k_B T \ln \frac{Q(T, M, N+1, at\ j)}{Q(T, M, N)}$$

$$= -k_B T \ln \left[\frac{(M-1)!(M-N)!}{(M-N-1)!M!} q_A \exp(-\beta U) \right]$$

$$= -k_B T \ln q_A + U - k_B T \ln(1-\theta) \qquad (5.49)$$

The placing of A at a fixed site involves three terms; first, creating a "cavity" at the jth site, second, introducing the internal Helmholtz energy of A, and finally the interaction between A and the site.

Combining Eq. (5.49) with Eq. (5.10) we get the chemical potential of the adsorbed molecules in the form

$$\mu_A^{ad} = \mu_A^* \left(at\ j\right) + k_B T \ln\theta \qquad (5.50)$$

This is the analogue of Eq. (4.30). Here the addition of one molecule to the system is split into two steps. First, we place the molecule at a specific site, say j. The corresponding work is μ_A^*, given in (5.49). Then we "liberate" the particle in the sense that the constraint on the fixed site is released. The corresponding change in the Helmholtz energy is $k_B T \ln\theta = k_B T \ln(N/M)$. In contrast to the case of an ideal gas (Eq. (4.30)) where we recovered the momentum PF, here the ligands have no translational degrees of freedom. The analog of $-k_B T \ln V$ in (4.30) is $-k_B T \ln M$, which results from the accessibility of all the M sites. As in Eq. (4.30), the term $k_B T \ln N$ arises from the assimilation of the released particles among the N indistinguishable particles. The term $k_B T \ln \theta$ is therefore the "*liberation*" *Helmholtz energy* of the adsorbed molecules.

The Helmholtz energy change for the process of bringing a molecule from a fixed position, say R_0, in an ideal-gas phase, to a fixed site, say j, is

$$\Delta A^*(R_0 \to site\ j) = \mu_A^*(at\ j) - \mu_A^*$$
$$= U - k_B T \ln(1 - \theta) \quad (5.51)$$

This quantity may be referred to as the *solvation* Helmholtz energy. The work involved in the "solvation" of a ligand consists of two terms. First, we have to create a "cavity" at a specific site, then we have the interaction between the ligand and the site.

For very "dilute" systems, i.e. when $\theta \ll 1$ we get for the solvation Helmholtz energy, entropy and energy the following results

$$\Delta A^* = U \quad (5.52)$$

$$\Delta S^* = -\frac{\partial \Delta A^*}{\partial T} = 0 \quad (5.53)$$

$$\Delta E^* = \Delta A^* + T \Delta S^* = U \quad (5.54)$$

In this particular example, the Helmholz energy of solvation is the same as the energy of solvation and the entropy of solvation is zero.

Adsorption of two kinds of ligands. See Fig. 5.5.

This is a simple generalization of the model discussed above; each site may adsorb one molecule; either a molecule of type A or a molecule of type B. Both A and B are at equilibrium with a reservoir at constant μ_A and μ_B,

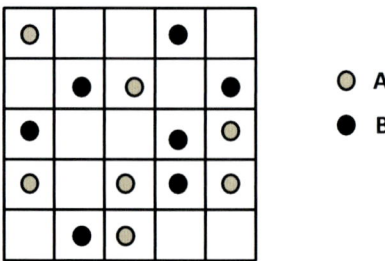

Fig. 5.5 Two kinds of ligands A and B.

respectively. This model is relevant to systems where two ligands compete for the same site. Since the sites are still independent we have the general result for the grand PF which is a straightforward generalization of Eq. (5.17). The details are left as an exercise. The PF is:

$$\Xi(T, M, \mu_A, \mu_B) = \xi^M \tag{5.55}$$

where ξ is the grand PF of a single site. For the case of a mixture of A and B, we have

$$\xi = q(0) + q(A)\lambda_A + q(B)\lambda_B \tag{5.56}$$

The three terms in (5.56) correspond to the three states of a single site; empty, occupied by A, and occupied by B, where the PF of a single site in the states i is $q(i)$. The average numbers of A and B molecules are

$$\bar{N}_A = \lambda_A M \frac{\partial \ln \xi}{\partial \lambda_A} = M \frac{q(A)\lambda_A}{1 + q(A)\lambda_A + q(B)\lambda_B} \tag{5.57}$$

$$\bar{N}_B = \lambda_B M \frac{\partial \ln \xi}{\partial \lambda_B} = M \frac{q(B)\lambda_B}{1 + q(A)\lambda_A + q(B)\lambda_B} \tag{5.58}$$

where we put $q(0) = 1$.

Note that in (5.56) we have

$$q(A) = q_A \exp\left(-\beta U_A\right) \tag{5.59}$$

$$q(B) = q_B \exp\left(-\beta U_B\right) \tag{5.60}$$

where U_A and U_B are the binding energies of A and B to the site. If either λ_A or λ_B is zero, then this case is reduced to the simple Langmuir model. Generalization to any number of ligands is straightforward.

5.6 Generalization to Two Kinds of Sites

We discuss here another simple generalization of the simple model discussed in Section 5.1. Here, we have two kinds of sites, say M_α sites of type A and M_β sites of type β, but only one type of ligand molecules A. See Fig. 5.6. The grand PF in this case is

$$\Xi(T, M_\alpha, M_\beta, \mu_A) = \xi_\alpha^{M_\alpha} \xi_\beta^{M_\beta} \tag{5.61}$$

where ξ_α is the grand PF of a single site:

$$\xi_\alpha = q_\alpha(0) + q_\alpha(1)\lambda_A \tag{5.62}$$

$$q_\alpha(1) = q_\alpha(0)q_\alpha \exp\left(-\beta U_\alpha\right) \tag{5.63}$$

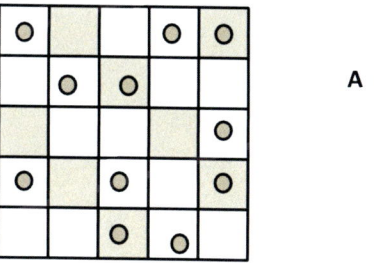

Fig. 5.6   Two kinds of sites.

Here, $q_\alpha(1)$ is the PF of a site of type α occupied by an A molecule with binding energy U_α and $q_\alpha(0)$ is the PF of an empty site of type α. Similar expressions apply for β.

The average number of adsorbed molecules in this case is

$$\bar{N}_A = \lambda_A \frac{\partial \ln \Xi}{\partial \lambda_A} = \lambda_A M_\alpha \frac{\partial \ln \xi_\alpha}{\partial \lambda_A} + \lambda_A M_\beta \frac{\partial \ln \xi_\beta}{\partial \lambda_A}$$

$$= M_\alpha \frac{q_\alpha(1)\lambda_A}{1 + q_\alpha(1)\lambda_A} + M_\beta \frac{q_\beta(1)\lambda_A}{1 + q_\beta(1)\lambda_A}$$

$$= \bar{N}_\alpha + \bar{N}_\beta \tag{5.64}$$

In this case, $\bar{N}_A$ is the sum of the average number of molecules adsorbed on sites of type α and on sites of type β. If either M_α or M_β is zero, this case is reduced to the simple Langmuir model.

We now calculate the analogs of the solvation quantities for this case. The reader is urged to work out the details. The chemical potential of the adsorbed ligand A, in a very dilute case ($\theta \ll 1$) is

$$\mu_A = k_B T \ln \theta - k_B T \ln q_A - k_B T \ln[x_\alpha q_\alpha + x_\beta q_\beta] \tag{5.65}$$

where q_α is the internal PF of the ligand A, and the quantities q_α and q_β are defined by

$$q_\alpha = \exp[-\beta U_\alpha], \quad q_\beta = \exp[-\beta U_\beta] \tag{5.66}$$

The analog of the pseudo-chemical potential for this case is

$$\mu_A^* = -k_B T \ln q_A - k_B T \ln[x_\alpha q_\alpha + x_\beta q_\beta] \tag{5.67}$$

The analog of the solvation quantities are

$$\Delta A^* = \mu_A^* - \mu_A^{ig} = -k_B T \ln[x_\alpha q_\alpha + x_\beta q_\beta]$$
$$= -k_B T \ln\langle \exp[-\beta B_A]\rangle_0 \qquad (5.68)$$

$$\Delta S^* = \frac{-\partial \Delta A^*}{\partial T} = -k_B \ln\langle \exp[-\beta B_A]\rangle_0 + \frac{\langle B_A\rangle_A}{T}$$
$$(5.69)$$

$$\Delta S^* = \langle B_A\rangle_A \qquad (5.70)$$

Here, we denoted by B_A the binding energy of A to the site. The average $\langle\ \rangle_0$ is defined in (5.68). This is an average with the probability distribution $x_\alpha = M_\alpha/M$ and $x_\beta = M_\beta/M$. Note carefully that this distribution does not depend on the binding energies of A to the different sites. x_α and x_β are the mole fractions for the two kinds of sites, and these are not affected by the binding process. On the other hand the quantity $\langle\ \rangle_A$ is defined as an average with the distribution.

$$y_\alpha = \frac{x_\alpha q_\alpha}{x_\alpha q_\alpha + x_\beta q_\beta}, \quad y_\beta = \frac{x_\beta q_\beta}{x_\alpha q_\alpha + x_\beta q_\beta} \qquad (5.71)$$

These probabilities depend on the binding energies. The different meanings of the two distributions are as follows. Before the ligand is adsorbed we have two kinds of sites. If we pick up a site at random, the probability that it is an α-site is x_α, and the probability that it is a β-site is x_β. On the other hand, y_α is the conditional probability that the site will be of type α, given that the site we picked up is occupied. Similarly, y_β is the probability that the site is β, given that it is occupied.

Note that ΔA^* depends only on the distribution (x_α, x_β). On the other hand, ΔE^* and ΔS^* also depend on the conditional distribution (y_α, y_β). This difference will be important in the system discussed in the next section.

5.7 Adsorption with Conformational Changes in the Adsorbent Molecules

In this section we discuss an important generalization of the model discussed in Section 5.1. We remove one of the assumptions, namely that the adsorbed ligand does not affect the adsorbent molecules. In many cases in biochemistry, the binding of the ligands causes a conformational change in the adsorbent molecules.

In the present section we deal with the simplest effect of a ligand on the conformational equilibrium of the adsorbent molecules. The study of this model is important for the understanding of cooperativity phenomena discussed in Chapter 6.

5.7.1 (***) *The Model and its Solution*

As in the simplest Langmuir model we start with M independent, identical and, for simplicity localized sites. Each site can adsorb only one ligand, which we assume to be a simple molecule A. The sites will be referred to as adsorbent molecules. It can be shown that all the results derived in this section are independent of whether the adsorbent molecules are free, or localized.[1]

The new feature of the present model is that each site or each polymer molecule can be in one of two states low energy (L) state, or high energy (H) state. These two states are in chemical equilibrium

$$L \rightleftarrows H \tag{5.72}$$

The adsorbed ligand A has different adsorption energies U_L and U_H, according to whether it is adsorbed on the L or the H form. The L and H forms can be thought of as two conformations of a polymer, which we symbolically denote by a square and a circle.

$$\square \; \rightleftarrows \; \bigcirc \tag{5.73}$$

The generalization of the simple Langmuir model applies to the two assumptions made in the introduction to Section 5.1. First, there are two adsorption parameters U_L and U_H instead of one. Second, the process of adsorption might perturb the sites, in the sense that the equilibrium composition of L and H might, in general be affected by the adsorption of A. Note that if L and H are not in equilibrium then we have two types of sites which are called L and H here, but otherwise it is the same model as the one discussed in Section 5.6. The new feature of this model is that changes in concentrations of L and H can occur.

In order to minimize the parameters needed to describe the model, we assume that L and H are characterized only by their *energy levels* E_L and E_H (with $E_L < E_H$). The PFs of a site in either state L or H are given by

$$Q_L = \exp\left(-\beta E_L\right), \quad Q_H = \exp\left(-\beta E_H\right) \tag{5.74}$$

Note that in general, E_L and E_H should be interpreted as free energies defined by

$$Q_L = \sum_{i \in L} \exp\left(-\beta E_i\right) = \exp\left(-\beta A_L\right) \tag{5.75}$$

where the sum $i \in L$ is over all states of the adsorbent molecules, which we recognize as belonging to the L state. The L state includes many microstates, which in this particular treatment we lump together in one state. From now on we refer to E_L and E_H as the only molecular parameters of the adsorbing molecules, and we suppress any other parameters that might enter into Q_L and Q_H (say, vibrations, rotations, and so on).

From the point of view of chemical equilibrium, we write the canonical PF of the system of M empty sites at temperature T as

$$Q(T, M) = (Q_L + Q_H)^M$$

$$= \sum_{M_L=0}^{M} \frac{M!}{M_L! M_H!} Q_L^{M_L} Q_H^{M_H}$$

$$= \sum_{M_L} Q^*(T, M_L, M_H) \qquad (5.76)$$

Note that if the molecules are in a gaseous phase, then Q is also a function of the volume of the system. In such a case we must include at least the translational PF of the molecules; i.e. instead of (5.76) we must have

$$Q(T, V, M) = \frac{V^M}{\Lambda^{3M} M!} (Q_L + Q_H)^M \qquad (5.77)$$

where V is the volume of the system and Λ^{3M} the momentum PF of the adsorbent molecules. Since the factor $V^M / \Lambda^{3M} M!$ is constant in the entire treatment that follows we can ignore it. (This would be the same as "freezing-in" the translation of the adsorbent molecules).

Following the discussion of chemical equilibrium in Section 4.4, we view Q^* defined in (5.76) as the PF of a "frozen-in" system with *fixed* values of M_L and M_H.

The equilibrium (or average) values of M_L and M_H are obtained from the condition

$$\frac{\partial \ln Q^*}{\partial M_L} = 0 \qquad (5.78)$$

which leads to the equilibrium condition

$$\frac{\bar{M}_H}{\bar{M}_L} = \frac{Q_H}{Q_L} \qquad (5.79)$$

We define the equilibrium constant for the empty sites, by

$$K = \frac{Q_H}{Q_L} = \exp[-\beta(E_H - E_L)] \qquad (5.80)$$

by our assumptions $E_H - E_L > 0$, and hence $0 \leq K \leq 1$. From (5.79) and (5.80) we have

$$\frac{\bar{M}_H}{\bar{M}_L} = K \qquad (5.81)$$

It should be noted that K is *defined* in (5.80) in terms of the *molecular* parameters E_H and E_L of the adsorbing molecules. By the equilibrium condition (5.79), it is also equal to the ratio $\bar{M}_H/\bar{M}_L$. However, in the following we will study the case where the equilibrium ratio $\bar{M}_H/\bar{M}_L$ changes (upon adsorption). In this case, the equality (5.81) does not hold, On the other hand, the meaning of K as an equilibrium constant defined by (5.80) remains unchanged.

We also introduce the mole fractions of sites in the two states by

$$x_L^0 = \frac{\bar{M}_L}{M} = \frac{Q_L}{Q_L + Q_H} = \frac{1}{1+K} \qquad (5.82)$$

$$x_H^0 = \frac{\bar{M}_H}{M} = \frac{Q_H}{Q_L + Q_H} = \frac{K}{1+K} \qquad (5.83)$$

These are also the probabilities of finding a specific site in either the L or the H state for the empty system. (The term empty refers here to the system of adsorbent molecules in the absence of ligands).

We now turn to the case where, in addition to the M sites, we also have N adsorbed ligands distributed over the sites. For simplicity, we assign no internal degrees of freedom to these molecules; they are characterized solely by their adsorption energies U_L and U_H.

The canonical partition function for such a system is

$$Q(T, M, N)$$

$$= \sum_{\substack{M_L + M_H = M \\ N_L + N_H = N}} \binom{M}{M_L} \binom{M_L}{N_L} \binom{M_H}{N_H} Q_L^{M_L} Q_H^{M_H} q_L^{N_L} q_H^{N_H}$$

$$(5.84)$$

where we used the notations

$$q_L = \exp(-\beta U_L), \quad q_H = \exp(-\beta U_H) \qquad (5.85)$$

N_L and N_H are the number of molecules adsorbed on L and H sites, respectively. The summation in (5.84) extends over all possible values of N_L, N_H, M_L, M_H with the conditions

$$M_L + M_H = M, \quad N_L + N_H = N,$$
$$N_L \leq M_L, \quad N_H \leq M_H \tag{5.86}$$

The auxiliary parameters N_L, N_H, M_L, M_H serve as intermediate quantities which determine the energy levels of the system, namely

$$E(N_L, N_H, M_L, M_H) = M_L E_L + M_H E_H$$
$$+ N_L U_L + N_H U_H \tag{5.87}$$

The degeneracy of these energy levels is given by the product of the three combinatorial factors in (5.84).

The summation in (5.84) cannot be carried out to obtain a closed form of the partition function. This is easily achieved however, by transforming to an open system with respect to the ligand molecules: The grand partition function for this system is

$$\Xi(T, M, \lambda)$$

$$= \sum_{N=0}^{M} \lambda^N Q(T, M, N)$$

$$= \sum_{M_L=0}^{M} \sum_{N_L=0}^{M_L} \sum_{N_H=0}^{M_H} \lambda^{N_L} \lambda^{N_H} \binom{M}{M_L} \binom{M_L}{N_L} \binom{M_H}{N_H}$$

$$\times Q_L^{M_L} Q_H^{M_H} q_L^{N_L} q_H^{N_H}$$

$$= \sum_{M_L=0}^{M} \binom{M}{M_L} Q_L^{M_L} (1 + \lambda q_L)^{M_L}$$

$$\times Q_H^{(M-M_L)}(1 + \lambda q_H)^{(M-M_L)}$$
$$= (Q_L + Q_H + \lambda q_L Q_L + \lambda q_H Q_H)^M = \xi^M$$

$$(5.88)$$

Here, $\lambda = \exp(\beta\mu)$ is the absolute activity of the ligand A, and μ is its chemical potential.

Note that $\Xi(T, M, \lambda)$ is the semi grand PF of a system of M adsorbent molecules and any number of ligands $\leq M$. No free ligands are taken into account in our definition of the system. Likewise, $Q(T, M, N)$ pertains to a system of M adsorbent and N ligand molecules.

As we have seen earlier in this chapter, the last form of the PF is typical of a system of independent sites; ξ, as defined in (5.88), may be viewed as the grand PF of a single site.

The four terms in ξ correspond to the four possible states of each site: empty L, empty H, occupied L, and occupied H. Symbolically, these states are denoted as

$$\square \quad \bigcirc \quad \boxed{\circ} \quad \textcircled{\circ} \qquad (5.89)$$

The terms in (5.88) correspond to the probabilities of finding a specific site in each of the states indicated in (5.89), namely

$$Pr(L, 0) = Q_L/\xi, \quad Pr(H, 0) = Q_H/\xi$$
$$Pr(L, A) = \lambda q_L Q_L/\xi, \quad Pr(H, A) = \lambda q_H Q_H/\xi \quad (5.90)$$

The probabilities of finding a site empty or occupied, independently of its state are

$$Pr(0) = Pr(L, 0) + Pr(H, 0) = \frac{Q_L + Q_H}{\xi} \qquad (5.91)$$

$$Pr(A) = Pr(L, A) + Pr(H, A)$$
$$= \frac{\lambda q_L Q_L + \lambda q_H Q_H}{\xi} \tag{5.92}$$

Note that as $\lambda \to 0$, $Pr(0) \to 1$, while $Pr(A) \to 0$. On the other hand, as $\lambda \to \infty$, $Pr(0) \to 0$, while $Pr(A) \to 1$. For later applications it will also be useful to introduce the conditional probability $Pr(L|A)$ of finding a site in state L when it is known to be occupied by a ligand A, i.e.

$$Pr(L|A) = \frac{Pr(L, G)}{Pr(A)} = \frac{\lambda q_L Q_L / \xi}{(\lambda q_L Q_L + \lambda q_H Q_H)/\xi}$$
$$= \frac{q_L Q_L}{q_L Q_L + q_H Q_H} \tag{5.93}$$

Similarly, the conditional probability of finding a site in state H, given that it is occupied is

$$Pr(H|A) = \frac{q_H Q_H}{q_L Q_L + q_H Q_H} \tag{5.94}$$

The average number of ligand molecules in the system is given by

$$\bar{N} = \lambda \frac{\partial ln \Xi}{\partial \lambda} = \frac{\lambda M}{\xi} (q_L Q_L + q_H Q_H) \tag{5.95}$$

Let θ be the average fraction of sites occupied. This is also equal to the probability of finding a specific site being occupied. From (5.90) and (5.95) we obtain

$$x = Pr(L, A) + Pr(H, A) = Pr(A) \tag{5.96}$$

By elimination of λ from (5.95), we obtain

$$\lambda = \left(\frac{\theta}{1-\theta}\right) \frac{Q_L + Q_H}{q_L Q_L + q_H Q_H} \tag{5.97}$$

From (5.97) we can obtain all the partial thermodynamic quantities of the adsorbed ligand. The chemical potential is

$$\mu = k_B T \ln \left(\frac{\theta}{1-\theta}\right) - k_B T \ln \left(\frac{q_L Q_L + q_H Q_H}{Q_L + Q_H}\right) \tag{5.98}$$

Using the distribution of two states in the empty site given in (5.82) and (5.83), we may rewrite (5.98) as

$$\mu = k_B T \ln \left(\frac{\theta}{1-\theta}\right) - k_B T \ln (q_L x_L^0 + q_H x_H^0)$$

$$= k_B T \ln \left(\frac{\theta}{1-\theta}\right) - k_B T \ln \langle \exp (-\beta B_A)\rangle_0 \tag{5.99}$$

where we have introduced the quantity B_A which will be referred to as the the *binding energy of A* to the site. More specifically, B_A is a function of the state of the site $\alpha = L, H$ and is defined as follows:

$$B_A(\alpha) = \begin{cases} U_L & \text{if } \alpha = L \\ U_H & \text{if } \alpha = H \end{cases} \tag{5.100}$$

In (5.99) we denoted $\langle \ \rangle_0$ an average of the function $\exp[-\beta B_A(\alpha)]$. The average is taken over all states of the adsorbent molecules, here L and H, with the probability distribution x_L^0 and x_H^0 as given in (5.82) and (5.83), i.e. of the empty system.

The partial molar (or molecular) entropy of the adsorbed ligand is obtained from (5.99) by differentiation with respect to the temperature:

$$\bar{S} = - \left(\frac{\partial \mu}{\partial T} \right)_{M,N}$$

$$= -k_B T \ln \left(\frac{\theta}{1-\theta} \right) + k_B \ln \langle \exp(-\beta B_A) \rangle_0$$

$$+ \frac{1}{T} \left[\frac{q_L Q_L (E_L + U_L) + q_H Q_H (E_H + U_H)}{q_L Q_L + q_H Q_H} \right.$$

$$\left. - \frac{q_L E_L + q_H E_H}{Q_L + Q_H} \right] \tag{5.101}$$

Using the conditional probabilities in (5.93) and (5.94), Eq. (5.101) may be written as

$$\bar{S} = -k_B \ln \left(\frac{\theta}{1-\theta} \right) + k_B \ln \langle \exp(-\beta B_A) \rangle_0$$

$$+ \frac{1}{T} [\langle E_P + B_A \rangle_A - \langle E_P \rangle_0] \tag{5.102}$$

E_P is the energy of the site or of the polymer, and is defined similarly to (5.100) as

$$E_p(\alpha) = \begin{cases} E_L & \text{for } \alpha = L \\ E_H & \text{for } \alpha = H \end{cases} \tag{5.103}$$

where the symbol $\langle \ \rangle_A$ signifies a *conditional* average taken with the distribution given in (5.93) and (5.94).

The partial molar (or molecular) energy of the ligand is given by

$$\bar{E} = \mu + T\bar{S} = \langle E_P + B_A \rangle_A - \langle E_P \rangle_0$$
$$= \langle B_A \rangle_A + (\langle E_P \rangle_A - \langle E_P \rangle_0) \tag{5.104}$$

It is important for the reader to work out the details of this example and make sure he or she understands the meaning of each average quantity.

5.7.2 The Adsorption Isotherm

From (5.97) we can solve for θ as a function of λ, i.e,

$$\theta = \frac{K_{in}\lambda}{1 + K_{in}\lambda} \tag{5.105}$$

where K_{in} is referred to as the *intrinsic* binding constant, and defined by

$$K_{in} = \frac{q_L Q_L + q_H Q_H}{Q_L + Q_H} = \langle \exp(-\beta B_A) \rangle_0. \tag{5.106}$$

Equation (5.105) is the adsorption isotherm, i.e. θ as a function in terms of the absolute activity λ of the ligand. We assume that the ligand is in equilibrium with an ideal gas phase, for which the chemical potential is

$$\mu_A^{ig} = k_B T \ln(\rho_A \Lambda_A^3) = k_B T \ln\left(\frac{P\Lambda_A^3}{k_B T}\right) \tag{5.107}$$

where P is the partial pressure of the ligand, given in (5.99). μ is the chemical potential of the adsorbed ligand, given in

(5.99). (In Section 5.2 this was denoted by μ^{ad}).

$$\mu_A^{ig} = \mu \tag{5.108}$$

For this special case we have

$$\theta = \frac{K_{ad}P}{1 + K_{ad}P} \tag{5.109}$$

where

$$K_{ad} = \frac{K_{in}\Lambda_A^3}{K_B T} \tag{5.110}$$

K_{ad} is the adsorption constant that can be obtained from the experimental curve of θ as a function of the pressure P.

The quantity denoted by K_{in} and defined in (5.105) will be referred to as the *intrinsic binding constant*. This quantity is related to the change in the Helmholtz energy for the process of bringing A from a fixed position, say R_0, in an ideal gas phase to a *fixed* and *empty* site, say j. The argument leading to this relation is similar to the one given in Section 5.4.

$$\Delta A^*(R_0 \rightarrow \text{ specific empty site } j)$$
$$= -k_B T \ln \frac{Q(T,M,N+1; \text{site } j \text{ occupied})}{Q(T,M,N; \text{empty site } j)}$$
$$= -k_B T \ln \left(\frac{q_L Q_L + q_H Q_H}{Q_L + Q_H} \right)$$
$$\quad - k_B T \ln \langle \exp\left(-\beta B_A\right)\rangle_0 \tag{5.111}$$

Comparing (5.111) with (5.106) we have the relation

$$K_{in} = \exp[-\beta\Delta A^*(R_0 \rightarrow \text{ specific empty site } j)] \tag{5.112}$$

We note that ΔA^* defined in (5.111) is different from a similar quantity defined for the process of bringing the ligand from R_0 to the specific site j (empty or occupied), for which we have the relation

$$\Delta A^* (R_0 \rightarrow \text{specific site } j) = \Delta A^* (R_0 \rightarrow \text{specific empty site } j)$$
$$- k_B T \ln (1 - \theta) \qquad (5.113)$$

6

Cooperativity in Binding Processes

Binding phenomena abounds in biological systems. The range of processes in which binding of ligands to specific sites on adsorbent molecules is immense: Oxygen molecules bind to specific sites on hemoglobin to be transported from point A to point B. Substrate molecules bind to specific site on enzymes before being converted to other molecules, proteins bind to a specific site on a specific gene to "decide" whether or not that gene will be expressed in the specific cell. This is, of course a tiny list of binding processes that occur in each of our cells.

In Chapter 5, we have studied the binding isotherm of a system of independent sites. Independence between sites means that binding of a ligand on one site does not affect the probability of binding of a second ligand on any other site. We have seen that independence among sites leads to the typical Langmuir binding isotherm. This is a modest, yet an important achievement of statistical thermodynamics.

In this chapter we will study the effect of *dependence* between sites on the binding isotherm. In doing so we will discover a common mechanism underlying many, seemingly unrelated, biochemical processes. Why is hemoglobin so *efficient* in transporting oxygen? How do some enzymes *regulate* and *maintain* the concentration of certain chemicals in our cells within narrow range? What makes the switching "on" and "off" of specific genes in specific enzymes?

The secret underlying all these processes will be referred to as *cooperativity*. The term "cooperativity" features in the literature in connection with many processes, having many, sometimes ill-defined meanings. In Section 6.1 we will define the term cooperativity in terms of probabilities of binding to different sites. In subsequent sections we will study the molecular *origin* of this cooperativity and its consequences in some real biochemical processes. We will also use the more colloquial language such as "communication" between ligands or "transmission" of information between ligands. The reader should carefully distinguish between the precise definition of the term cooperativity, and the more loosely, yet sometimes useful descriptions of this term.

In this chapter, we will always consider a system consisting of M adsorbent molecules. Each molecule is denoted by P, and contains M sites on which ligand molecules A can bind. The whole system is at constant temperature T, and volume V and total number of ligands N. In some cases it will be convenient to use an open system with respect to the ligands. We will also assume that the adsorbent molecules are *localized*, i.e. they do not have translational degrees of freedom. This assumption is made for simplicity. It does not affect the results.[1]

6.1 The General Definition of Cooperativity in a Two-Site System

Our system is a single adsorbent molecule denoted P, each having two binding sites denoted by a and b. These sites could be identical or different, but in this section the treatment is general and can be applied to any case.

We will use probabilistic language to define the terms "correlation" and "cooperativity." As a rule, the probabilities pertaining to the various events are read from the appropriate terms in the grand PF of a single adsorbent molecule, which is referred to as the system. In the present case the PF of a single adsorbed molecule is

$$\xi(T,\lambda) = Q(0,0) + [Q(a,0) + Q(0,b)]\lambda + Q(a,b)\lambda^2$$

$$(6.1)$$

The four terms in Eq. (6.1) correspond to the four states of the system, Fig. 6.1. These are: both sites empty, site a occupied, site b occupied, and both sites occupied. The coefficients Q in Eq. (6.1) are the PF of a single system in the various states shown in Fig. 6.1.

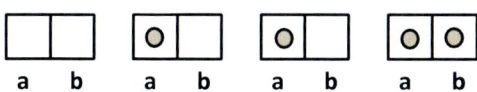

Fig. 6.1 The four occupancy states of a two-site system.

The average number of ligands *per system* is

$$n = \lambda \frac{\partial \ln \xi}{\partial \lambda}$$

$$= \frac{[Q(a,0) + Q(0,b)]\lambda + 2Q(a,b)\lambda^2}{Q(0,0) + [Q(a,0) + Q(0,b)]\lambda + Q(a,b)\lambda^2} \qquad (6.2)$$

The average number of ligands per *site* is therefore

$$\theta = \frac{1}{2}n \qquad (6.3)$$

where λ is the absolute activity of the ligand. We will always assume that the ligand is at equilibrium with a reservoir which is maintained at a constant chemical potential μ, or equivalently at a constant absolute activity λ. The relationship between the two is $\mu = k_B T \ln \lambda$. We will also assume that the ligand in the reservoir is either an ideal gas or an ideal dilute solution, in both cases λ is proportional to the concentration (C) of the ligand, i.e. $\lambda = \lambda_0 C$. Note that λ is a dimensionless quantity. On the other hand, λ_0 must have the dimensions of C^{-1} such that the product $\lambda_0 C$ is a dimensionless number.

We now define the three *intrinsic* binding constant to the site a, b and ab by

$$k_a = \frac{Q(a,0)}{Q(0,0)}\lambda_0, \quad k_b = \frac{Q(0,b)}{Q(0,0)}\lambda_0 \text{ and } k_{ab} = \frac{Q(a,b)}{Q(0,0)}\lambda_0^2$$

$$(6.4)$$

In terms of the intrinsic binding constants, we write the binding isotherm, $\theta(C)$ as

$$\theta = \frac{1}{2}\frac{(k_a + k_b)C + 2k_{ab}C^2}{1 + (k_a + k_b)C + k_{ab}C^2} \qquad (6.5)$$

The four fundamental probabilities, corresponding to the four "events" depicted in Fig. 6.1 may be read directly from the PF in Eq. (6.1). These are

$$Pr(0,0) = Q(0,0)/\xi, \quad Pr(a,0) = Q(a,0)\lambda/\xi$$
$$Pr(0,b) = Q(0,b)\lambda/\xi, \quad Pr(a,b) = Q(a,b)\lambda^2/\xi \tag{6.6}$$

Note that the sum of these probabilities is unity.

We will also need the marginal probabilities. These are defined by:

$$Pr(a) = Pr(a,\_) = Pr(a,0) + Pr(a,b)$$
$$Pr(b) = Pr(\_,b) = Pr(0,b) + Pr(a,b) \tag{6.7}$$

A blank space in the argument of Pr means "unspecified" or "anything." Thus, $Pr(a,\_)$, [or simply $Pr(a)$] is the probability of the event "site a is occupied." The state of the site b is *unspecified*; it could be either empty or occupied. A similar meaning applies to $Pr(b)$.

The *correlation* between *any* two events $\mathcal{A}$ and $\mathcal{B}$ is defined by

$$g(\mathcal{A}, \mathcal{B}) = \frac{Pr(\mathcal{A} \cdot \mathcal{B})}{Pr(\mathcal{A}) \times Pr(\mathcal{B})} \tag{6.8}$$

Clearly, this quantity measures the extent of *dependence* between the two events. The two events $\mathcal{A}$ and $\mathcal{B}$ are said to be independent if and only if

$$Pr(\mathcal{A} \cdot \mathcal{B}) = Pr(\mathcal{A}) \times Pr(\mathcal{B}) \tag{6.9}$$

or, equivalently

$$g(\mathcal{A}, \mathcal{B}) = 1 \tag{6.10}$$

In the context of the theory of probability the term *correlation* as defined in Eq. (6.8) applies to *any* two events. For instance, the correlation between the events "site a is empty" and "site b is empty" is

$$g(0,0) = \frac{Pr(0,0)}{Pr(0,\_)Pr(\_,0)} \qquad (6.11)$$

The same applies for the two events "site a is occupied" and "site b is occupied", for which the correlation is

$$g(a,b) = \frac{Pr(a,b)}{Pr(a)P(b)} = \frac{Pr(a,b)}{Pr(a,\_)Pr(\_,b)} \qquad (6.12)$$

The term *correlation* as a measure of the extent of the dependence can be applied to *any* two events. On the other hand, the term *cooperativity* is applied only to a subclass of events involving ligands occupying sites. It will measure the extent of the dependence between two (or more) *ligands* bound to their sites. For such events, the terms *cooperativity* and *correlation* will be used synonymously. Whenever events of this type are correlated, we will say that the ligands occupying the two sites *cooperate*, or simply that the system is *cooperative*. Thus, the binding is said to be cooperative whenever binding on one site affects the probability of binding on another site. Sometimes we will also say loosely that in this case there exists *communication* between the two ligands occupying the sites a and b. Communication is used here in the sense that a ligand at one site "knows" or "senses" the presence of another ligand at the second site.

Whenever $g(a, b) = 1$, we say that there is no correlation between the events "site a is occupied" and "site b is occupied." Such a system is said to be *non-cooperative*. If $g(a, b) > 1$, we say that the two ligands cooperate *positively*, and when $g(a, b) < 1$, they cooperate *negatively*. In some earlier publications the term "cooperativity" is used for *positive cooperativity* and "anti-cooperativity" is used for *negative cooperativity*. In this book "cooperativity" is used whenever $g \neq 1$.

In terms of conditional probabilities we have the following relationships

$$Pr(a|b) = \frac{Pr(a, b)}{Pr(b)} = Pr(a)g(a, b) \quad \text{and}$$

$$Pr(b|a) = \frac{Pr(a, b)}{Pr(a)} = Pr(b)g(a, b)$$

(6.13)

Thus, positive cooperativity means that the conditional probability, say $Pr(a|b)$, is larger than the unconditional probability, i.e. the fact that "b is occupied" *enhances* the probability of the event "a is occupied." Sometimes, when there is positive cooperativity one says that a ligand at a *supports* or *favors* the binding of a ligand at b, and vice versa. Since $g(a, b)$ is defined symmetrically with respect to a and b, it follows from $g(a, b) > 0$ that both $Pr(a, b) > P(a)$ and $Pr(b|a) > Pr(b)$. Similarly, negative cooperativity, $g(a, b) < 1$, implies that $Pr(a|b) < Pr(a)$, and $Pr(b|a) < Pr(b)$.

In general the extent of correlation, $g(a, b)$ is dependent on λ. Two limiting cases are of interest:

$$\lim_{\lambda \to \infty} g(a, b) = 1$$

(6.14)

and

$$g^0(a, b) \equiv \lim_{\lambda \to 0} g(a, b) = \frac{Q(a, b)Q(0, 0)}{Q(a, 0)Q(0, b)} \qquad (6.15)$$

Note carefully that when $\lambda \to \infty$, each of the probabilities $Pr(a)$, $Pr(b)$ and $Pr(a, b)$ becomes unity, simply because all sites are occupied. Hence, also $g(a, b) \to 1$. On the other hand, when $\lambda \to 0$, each of these probabilities tend to *zero*, but the ratio defining $g(a, b)$ in Eq. (6.11) tends to a constant.

Exercise: Use the probabilities defined in (6.6) and the correlation in (6.11) to derive the result (6.15).

Figure 6.2 shows the dependence of $g(a, b)$ on λ for various values of $g^0(a, b)$. Whatever the initial value of $g^0(a, b)$, the correlation function $g(a, b)$ will tend to unity when $\lambda \to \infty$. The significance of $g^0(a, b)$ will become clear in the following sections.

In all our applications in this chapter we will need only $g^0(a, b)$. This is the only quantity that appears in the binding isotherm. For instance, if we wish to express k_{ab} in Eq. (6.4) in terms of k_a and k_b, we find that

$$k_{ab} = \frac{Q(a, b)}{Q(0, 0)} \lambda_0^2 = g^0(a, b)k_a k_b \qquad (6.16)$$

For this reason, we will always refer to $g^0(a, b)$ as the *pair correlation* or the *cooperativity* in the system, and drop the subscript zero.

The concept of cooperativity can also be interpreted in terms of Gibbs or Helmholtz energy change. We define

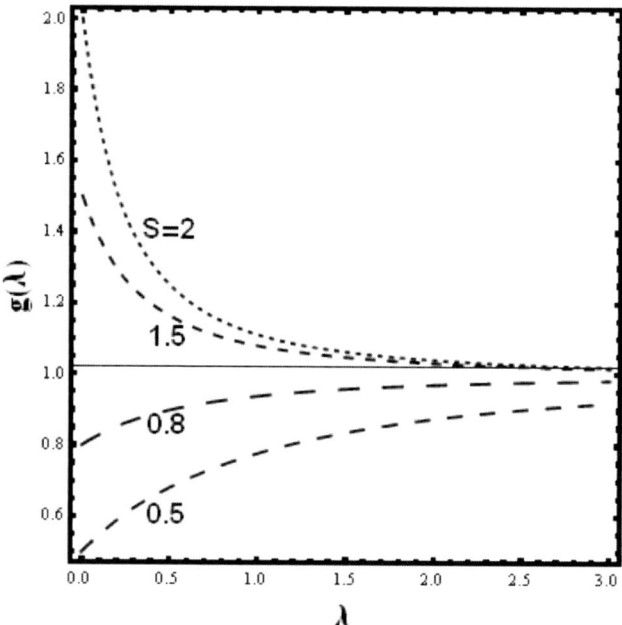

Fig. 6.2 The dependence of $g(a, b)$, defined in Eq. (6.12), on λ for the special case $Q(a, b) = S$, $Q(a, 0) = Q(0, b) = Q(0, 0) = 1$. This is the case of *direct* correlation only, discussed in Section 4.3. The values of the initial correlation $g(\lambda, 0) = g^0(a, b) = S$ are indicated next to each curve.

the quantity

$$W(a, b) = -k_B T \ln g(a, b) = -k_B T \ln \frac{Q(a, b)Q(0, 0)}{Q(a, 0)Q(0, b)}$$

(6.17)

where $W(a, b)$ is the work involved in the "reaction" that may be written symbolically as

$$(a, 0) + (0, b) \rightarrow (a, b) + (0, 0)$$

(6.18)

i.e. the formation of a fully occupied system (a, b) from two singly occupied systems.

The quantity $W(a, b)$ is the analog of the *potential of mean force* (PMF) in the theory of liquids. We will see in the following sections that this quantity, sometimes behaves as *energy*, but in most cases it is a *free energy*. In the theory of liquids $W(r)$ is the work associated with bringing two particles from infinite separation to the final distance r. If there are no other (solvent) molecules, $W(r)$ is simply the potential energy of interaction denoted $U(r)$. We will see the analog of both cases in the succeeding sections. In terms of $W(a, b)$ we may also define non-cooperative systems whenever $W(a, b) = 0$, and positive and negative cooperativity for $W(a, b) < 0$ and $W(a, b) > 0$, respectively. The reader should note the potentially confusing statement that a *positive* cooperativity involves *negative* values of $W(a, b)$, and vice versa.

6.2 Direct Correlation Between Two Identical Sites

The simplest and most obvious case of cooperativity occurs when there exists *direct* interaction between the two ligands occupying the two sites. The internal properties of the adsorbent molecule are still assumed to be unaffected by the binding process. The four possible states of the system are as in Fig. 6.1, but in our case a and b are identical. The corresponding coefficients in the Grand PF are:

$$Q(0, 0), \quad Q(1, 0) = Q(0, 1) = Q(0, 0)q,$$
$$Q(1, 1) = Q(0, 0)q^2 S \tag{6.19}$$

where

$$q = \exp(-\beta U), \quad S = \exp[-\beta U(1, 1)] \tag{6.20}$$

U is the binding *energy* of the ligand to the site. $U(1, 1)$ is the *direct* interaction energy between the two ligands. By *direct* we mean the intermolecular interaction energy between the two ligands as if they were in a *vacuum*.

The binding isotherm for this system is

$$\theta = \frac{n}{2} = \frac{q\lambda + q^2 S\lambda^2}{1 + 2q\lambda + q^2 S\lambda^2} = \frac{kC + k^2 SC^2}{1 + 2kC + k^2 SC^2}$$

$$(6.21)$$

where the *intrinsic* binding constant is $k = q\lambda_0$. In this particular model we identify the correlation function as:

$$g(1, 1) = \frac{Q(1, 1)Q(0, 0)}{Q(1, 0)Q(0, 1)} = S = \exp[-\beta U(1, 1)] \quad (6.22)$$

and the corresponding work, see Eq. (6.17)

$$W(1, 1) = -k_B T \ln g(1, 1) = U(1, 1) \qquad (6.23)$$

Note that in the case of two identical sites, occupied and empty sites are denoted by 1 and 0, respectively. When the sites are different we replace the "1" by the symbol used to name the site, say "*a*" or "*b*," and insert zero when the site is empty.

We will refer to this particular correlation as *direct* correlation, simply because it originates from the *direct* interaction between the two ligands. These interactions could be quite strong, as in the case of two protons in dicarboxylic acids (Section 6.5.5). In other cases they are very weak and can be neglected. Such is the case for oxygen molecules occupying the sites of hemoglobin. For a long time the cooperativity in binding of oxygen to hemoglobin was a mystery. Because of the large distances between the oxygen molecules, it was clear that the *direct* interaction between the oxygen molecules could not be responsible for the cooperativity. We will discuss the source of such, *indirect* correlations, hence cooperativity in the following sections.

In the present system, $S = 1$, [or equivalently, $U(1, 1) = 0$] means that the system is non-cooperative. The system is positively, or negatively cooperative when $S > 1$, or $S < 1$, respectively.

The effect of $U(1, 1)$ on the conditional probabilities [see Eq. (6.13)] is easily understood. When attraction exists between the ligands, a ligand occupying one site *attracts* the second ligand, and hence increases the probability of binding the second ligand, hence positive cooperativity. The reverse holds when the two ligands repel each other.

Two properties of the *direct* correlation should be noted, especially when compared with indirect correlation. (1) The interaction energy $U(1, 1)$, and hence $g(1, 1)$ are strongly dependent on the distance between the ligands. (2) Since $U(1, 1)$ is temperature-independent, the correlation $g(1, 1)$ depends on T only through $\beta = (k_B T)^{-1}$.

Figure 6.3 shows a few binding isotherms $\theta(x)$, where $x = q\lambda = kC$ is a dimensionless concentration, for (a) $S \geq 1$, and (b) $S \leq 1$. Note that the initial slope of $\theta(C)$ is determined by k. The initial curvature is determined by both k and S,

$$\left(\frac{\partial \theta}{\partial C} \right) \Big|_{C=0} = k, \quad \left(\frac{\partial^2 \theta}{\partial C^2} \right) \Big|_{C=0} = 2k^2(S - 2) \quad (6.24)$$

When $S < 1$, i.e. negative cooperativity, the binding isotherm starts with a negative curvature [see Eq. (6.24)].

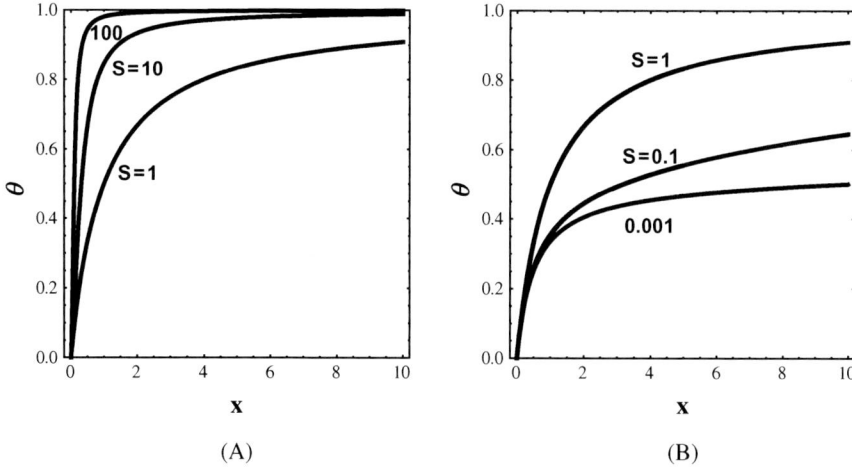

(A)   (B)

Fig. 6.3 The binding isotherm for (A) positive cooperativity with $S = 1, 10\ 100$, and (B) negative cooperativity with $S = 0.1, 0.001, 1$. Values of i are indicated next to each curve.

When S is very small, we observe an apparent "saturation" at $\theta = 1/2$, but at higher values of x we reach the eventual saturation at $\theta = 1$. This is difficult to observe in a plot of $\theta(x)$, as in Fig. 6.3. It is therefore more convenient to plot θ as a function of $y = \log_{10} x$, where we can plot the binding isotherm for both positive and negative cooperativities, with a much wider range of values S. One should be careful about the double meaning of the symbol θ; once as the *name* of the function, and once as the *value* of the fraction of sites occupied. In writing $\theta(\log_{10}(x))$ we do not refer to the compound function $f(x) = \theta(\log_{10}(x))$, but to the *values* of θ plotted as a function of the new variable $y = \log_{10}(x)$. The actual function plotted is $g(y) = \theta(10^y)$, i.e. it is obtained from the original binding isotherm, $\theta(x)$ by substituting $x \to 10^y$. The choice of $y = \log_{10} x$ is convenient because it allows a large variation in x, and it also conforms

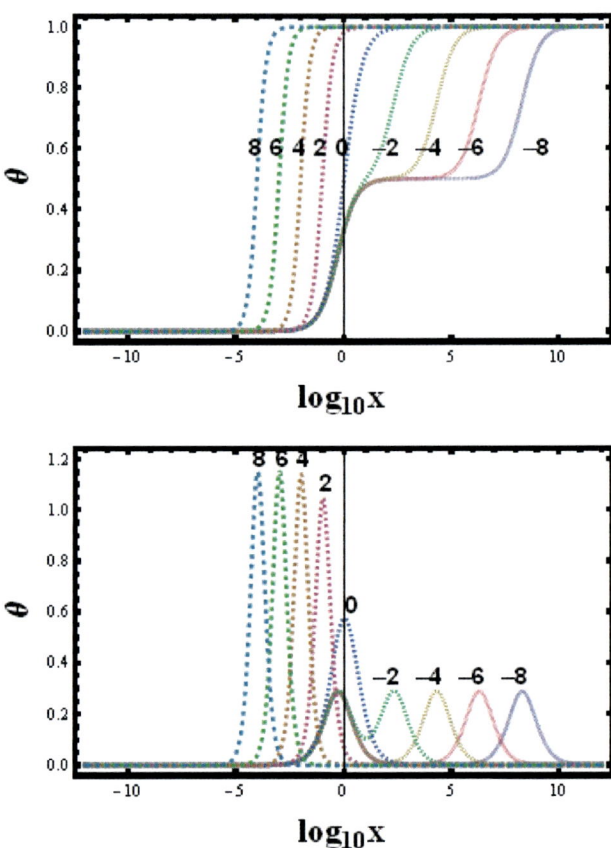

Fig. 6.4 θ as a function of $\log_{10} x$ for $S = 10^i (-8 \le i \le 8)$. The plotted function is $f(y) = \theta(10^y)$, where we substitute $x \to 10^y$ in $\theta(x)$. Below are the derivatives, $f'(y)$, plotted as a function of $y = \log_{10} x$. Values of i are indicated next to each curve.

to the tradition of plotting titration curves, where $pH = -\log_{10}[H]$. Figure 6.4 shows $\theta(\log_{10} x)$ for $S = 10^i$, with $-8 \le i \le 8$, and the corresponding derivatives. It is clearly seen that both positive and negative cooperativities can be exhibited in one plot. Also, the apparent saturation at $\theta = 1/2$ and the eventual saturation at $\theta = 1$ as $x \to \infty$ can be seen in the same plot. These plots are also closely related to the titration curves, where we choose the pH scale for the abscissa.[2] Note that the $S = 0$ curve has a single point at $x = 0$, at which the slope is maximal.

6.3 (***) Two Different Sites; Genuine and Spurious Cooperativity

The model discussed in this section is essentially the same as in the previous section, except that now we have two *different* sites. The corresponding coefficients of the PF are

$$Q(0,0), \quad Q(a,0) = Q(0,0)q_a,$$
$$Q(0,b) = Q(0,0)q_b, \quad Q(a,b) = Q(0,0)q_a q_b S_{ab} \tag{6.25}$$

where

$$q_a = \exp(-\beta U_a), \quad q_b = \exp(-\beta U_b),$$
$$S_{ab} = \exp[-\beta U(a,b)] \tag{6.26}$$

The binding isotherm for this case is [compare with Eq. (6.21)]

$$\theta = \frac{n}{2} = \frac{1}{2}\frac{(q_a + q_b)\lambda + 2q_a q_b S_{ab}\lambda^2}{1 + (q_a + q_b)\lambda + q_a q_b S_{ab}\lambda^2}$$
$$= \frac{1}{2}\frac{(k_a + k_b)C + 2k_a k_b S_{ab}C^2}{1 + (k_a + k_b)C + k_a k_b S_{ab}C^2} \tag{6.27}$$

where $k_a = q_a \lambda_0$ is the intrinsic binding constant for the site α. Note that the binding isotherm in this model is determined by *three* constants, k_a, k_b and S_{ab} as compared with only two in the model of Section 6.2. Experimentally, only the two quantities K_1 and K_2 defined by

$$K_1 = k_a + k_b, \quad K_1 K_2 = k_a k_b S_{ab} \tag{6.28}$$

may be determined from the binding isotherm in Eq. (6.27).

Clearly, the three quantities k_a, k_b and S_{ab} cannot be determined from the two experimental quantities K_1 and K_2. To obtain all three parameters k_a, k_b and S_{ab} one can measure the *individual* binding isotherms for the two sites. The latter are defined as the average occupation number of ligands at the specific site. Thus,

$$\theta_a = n_a = Pr(a, 0) + Pr(a, b) = \frac{Q(a, 0)\lambda}{\xi} + \frac{Q(a, b)\lambda^2}{\xi}$$

$$= \frac{k_a C + k_a k_b S_{ab} C^2}{1 + (k_a + k_b)C + k_a k_b S_{ab} C^2} \tag{6.29}$$

and similarly

$$\theta_b = \frac{k_b C + k_a k_b S_{ab} C^2}{1 + (k_a + k_b)C + k_a k_b S_{ab} C^2} \tag{6.30}$$

It is important to note that θ_a is not the binding isotherm of an *isolated* site of type a. To see this we rewrite θ_a in Eq. (6.29) in the modified form

$$\theta_a = \frac{k_a^* C}{1 + k_a^* C} \quad \text{with } k_a^* = k_a \frac{1 + k_b S_{ab} C}{1 + k_b C} \tag{6.31}$$

and

$$\theta = \theta_a + \theta_b \tag{6.32}$$

Thus, although each of θ_a and θ_b is a simple Langmuir isotherm, the average of the two has a different form.

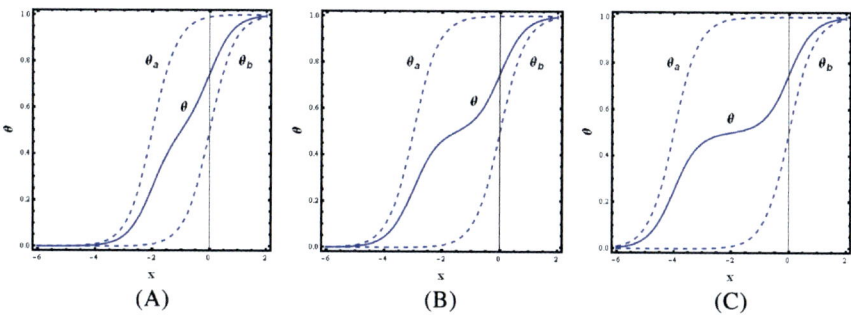

Fig. 6.5 The binding isotherms for a two-site system with different binding constants: (A) $k_a = 1$, $k_b = 100$; (B) $k_a = 1$, $k_b = 10^3$; (C) $k_a = 1$, $k_b = 10^4$. The full line is θ [see Eq. (6.32)]. The dashed lines are θ_a and θ_b.

Figure 6.5 shows θ_a, θ_b and θ for a *non-cooperative* ($S_{ab} = 1$) system with $k_a = 1$ and $k_b = 10^2, 10^3$ and 10^4.

A glance at Fig. 6.5 shows that the binding isotherm of the system is similar to the binding curve of a system with negative cooperativity (Fig. 6.4). In order to understand the source of this similarity in the binding isotherm consider the following three systems; Fig. 6.6.

(A) A system of 2M independent molecules each having a single site, $M_L = M$ of which are of type L and $M_H = M$ of type H. We use the notation L and H for a and b, respectively to avoid confusion with the three cases (A), (B) and (C) shown in Fig. 6.5. The corresponding grand PF is:

$$\Xi_A = \frac{\xi_L^{M_L} \xi_H^{M_H}}{M_L! M_H!}$$

$$= \frac{Q_L(0)^{M_L}(1 + q_L\lambda)^{M_L} Q_H(0)^{M_H}(1 + q_H\lambda)^{M_H}}{M_L! M_H!}$$

$$(6.33)$$

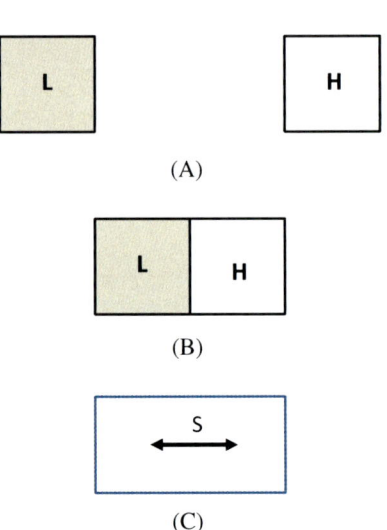

(A)

(B)

(C)

Fig. 6.6 Schematic illustration of the three equivalent systems **A**, **B**, and **C**, corresponding to Eqs. (6.33), (6.34) and (6.36). System (**A**) consists of a mixture of single-site particles, in (**B**) we have double-site particles where the sites are *different* and *independent* and in (**C**) the two sites are *identical* but *dependent*.

(**B**) A system of M *identical and independent* two-site molecules each having two *different* and *independent* sites. The corresponding grand PF is:

$$\Xi_B = \frac{Q(0,0)^M (1 + (q_L + q_H)\lambda + q_L q_H \lambda^2)^M}{M!} \quad (6.34)$$

Clearly, the two systems (**A**) and (**B**) are different. Note however that the factors $Q_L(0)$, $Q_H(0)$ and $Q(0,0)$ and the factorials $M_L!$, $M_H!$ and $M!$ do not affect the binding isotherm. Therefore, when $M_L = M$ and $M_H = M$, the two systems are *equivalent* in the sense of having the same binding isotherm. This follows immediately from the identity of the two polynomials

$$(1 + q_L \lambda)(1 + q_H \lambda) = 1 + (q_L + q_H)\lambda + q_L q_H \lambda^2$$

$$(6.35)$$

The more remarkable fact is that the two systems (A) and (B) are also equivalent (in the sense of having the same binding isotherms) to another system:

(C) A system of M identical and independent adsorbent molecules, each having two *identical* but *dependent* sites, with one intrinsic constant k and pair correlation S. The grand partition function of this system is:

$$\Xi_C = \frac{Q(0,0)^M (1 + 2q\lambda + q^2 S \lambda^2)^M}{M!} \qquad (6.36)$$

The two systems (A) and (C) are equivalent provided that

$$q = \frac{q_L + q_H}{2}, \quad S = \frac{4q_L q_H}{(q_L + q_H)^2} \qquad (6.37)$$

This is equivalent to the conditions,

$$k = \frac{k_L + k_H}{2}, \quad S = \frac{4k_L k_H}{(k_L + k_H)^2} \qquad (6.38)$$

The corresponding binding isotherm for this case is

$$\theta_C = \frac{kC + k^2 S C^2}{1 + 2kC + k^2 S C^2} \qquad (6.39)$$

From the experimental binding isotherm one cannot distinguish between the three systems (A), (B) and (C). (Note however, that other properties of the system, such as the pressure, energy, entropy, etc. are different). This fact could, in some cases, lead to misinterpretation of the cooperative behavior of the system. If $k_L \neq k_H$, then S defined as in Eq. (6.38) must be smaller than unity. This follows from the identity

$$S = \frac{4k_L k_H}{(k_L + k_H)^2} = 1 - \frac{(k_L - k_H)^2}{(k_L + k_H)^2} \leq 1 \qquad (6.40)$$

i.e. the binding isotherm of system (C) is negatively cooperative.

Suppose that we know that our system is either (A) or (**B**), then observing a binding isotherm of the form of Figure 6.5 is unlikely to mislead us into believing that our system is cooperative. The two peaks in the slope-plot will be recognized as originating from the two *different* binding constants k_L and k_H. Therefore, the apparent cooperativity observed in either (A) or (B) must be *spurious*. It will be distinguished from the genuine cooperativity when the system is (C). Of course, if we observe a binding isotherm of the type of Fig. 6.5 without knowing the system, we cannot tell whether the system is cooperative or not.

The qualitative physical reason for the equivalence of the systems (A) (or (B)) and (C) is simple. Consider the case where $k_H \gg k_L$, in which case we find an apparent "saturation" at $\theta = 1/2$. The (negative) spurious cooperativity in this case is

$$S = \frac{4k_L k_H}{(k_L + k_H)^2} \approx 4\frac{q_L}{q_H} \ll 1 \qquad (6.41)$$

A system (C), with very strong (genuine) negative cooperativity, will *first* be filled only at one of its sites on each molecule. The reason for this behavior is that two ligands on the same molecule will strongly repel each other. Only at very high ligand activity will the ligands be forced to occupy the *second* site on the molecules. When the system is either (A) or (B), then again, because $k_H \gg k_L$, the sites of type H will be filled first, and sites of type L will remain empty, not because of *repulsion*, but simply because of the overwhelming preference to binding on H. Again, we will observe an apparent "saturation" at $\theta = 1/2$. At very high ligand activities, when nearly all the preferred sites H have been occupied, the sites of type L will start to be occupied. Thus, we see that the patterns of occupation, and hence the binding isotherm, are the same although the physical reasons for the apparent saturation at $\theta = 1/2$ are different.

The main moral of this section is that cooperativity *cannot* be determined only from the *shape* of the binding isotherm.

6.4 Indirect Correlations Due to the Conformational Changes Induced by the Ligands

The model described in this section is an extension of the model of Section 6.2. Here, the adsorbent molecule has two macrostates, and a transition between the two macrostates may be induced by binding of a ligand. As in Section 6.2, the adsorbent molecule has two sites which can either be identical or different.

We start with two identical sites. Since the sites are distinguishable (the system is localized), there are altogether eight states for the system, shown in Fig. 6.7. The grand partition function of an adsorbent molecule is

$$\xi = Q(0) + Q(1)\lambda + Q(1)\lambda^2 \qquad (6.42)$$

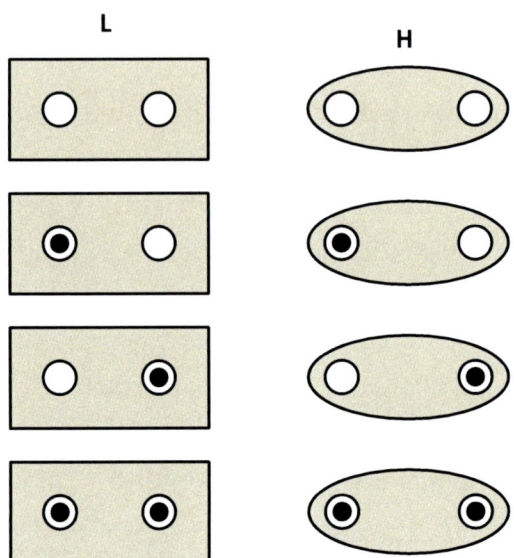

Fig. 6.7 The eight possible configurations of a two-state system with two identical sites. The L form is represented by a rectangle (white) and the H form by an eclipse (black).

where, we used the notations

$$Q(0) = Q(L; 0, 0) + Q(H; 0, 0)$$
$$Q(1) = Q(L; 1, 0) + Q(H; 1, 0) + Q(L; 0, 1) + Q(H; 0, 1)$$
$$Q(2) = Q(L; 1, 1) + Q(H; 1, 1)$$

$$(6.43)$$

and

$$q_L = \exp(-\beta U_L), \quad q_H = \exp(-\beta U_H)$$
$$S_L = \exp[-\beta U_L(1, 1)], \quad S_L = \exp[-\beta U_H(1, 1)]$$

$$(6.44)$$

Using these notations, the binding isotherm has the form

$$
\begin{aligned}
\theta = \frac{\bar{n}}{2} \\
= \frac{(Q_L q_L + Q_H q_H)\lambda + (Q_L q_L^2 S_L + Q_H q_H^2 S_H)\lambda^2}{(Q_L + Q_H) + 2(Q_L q_L + Q_H q_H)\lambda + (Q_L q_L^2 S_L + Q_H q_H^2 S_H)\lambda^2} \\
= \frac{k_1 C + k_{11} C^2}{1 + 2k_1 C + k_{11} C^2}
\end{aligned} \tag{6.45}
$$

The two intrinsic constants k_1 and k_{11} are defined in (6.45). We will need the probabilities of the following events:

$$
Pr(L; 0, 0) = \frac{Q(L; 0, 0)}{\xi}, \quad Pr(L; 1, 0) = \frac{Q(L; 1, 0)\lambda}{\xi},
$$
$$
Pr(L; 1, 1) = \frac{Q(L; 1, 1)\lambda^2}{\xi} \tag{6.46}
$$

Here, $Pr(L; 0, 0)$ is the probability of finding the adsorbent molecule in state L and empty. A similar meaning applies to the probabilities $Pr(L; 1, 0)$ and $Pr(L; 1, 1)$, etc.
The marginal probabilities are

$$
x_L = Pr(L) = (Q_L + 2Q_L q_L \lambda + Q_L q_L^2 S_L \lambda^2)/\xi
$$
$$
x_H = Pr(H) = (Q_H + 2Q_H q_H \lambda + Q_L q_H^2 S_H \lambda^2)/\xi \tag{6.47}
$$

Here, x_L and x_H are the mole fractions, or the probabilities of finding the conformation L and H, respectively. Three special cases are:

(i) The empty system: In this case

$$x_L^0 = \frac{Pr(L; 0, 0)}{Pr(0, 0)} = \frac{Q_L/\xi}{(Q_L + Q_H)/\xi}$$

$$= \frac{Q_L}{Q_L + Q_H}, \quad x_H^0 = 1 - x_L^0 \qquad (6.48)$$

which is the conditional probability of finding L given that the system is empty. This is the same as the equilibrium mole fraction of the empty adsorbent molecule.

(ii) The singly occupied system: The conditional probability of finding L given a specific site (say the lhs) occupied and the second site empty, is

$$x_L^{(1)} = \frac{Pr(L; 1, 0)}{Pr(1, 0)} = \frac{Q_L q_L \lambda/\xi}{(Q_L q_H + Q_H q_H)\lambda/\xi}$$

$$= \frac{Q_L q_L}{Q_L q_L + Q_H q_H} \qquad (6.49)$$

(When the sites are different, say a and b, we must distinguish between $x_L^{(a)}$ and $x_L^{(b)}$).

(iii) The doubly occupied system: The conditional probability of finding L given that the system is doubly occupied is

$$x_L^{(2)} = \frac{Pr(L; 1, 1)}{Pr(1, 1)} = \frac{Q_L q_L^2 S_L}{Q_L q_L^2 S_L + Q_H q_H^2 S_H} \qquad (6.50)$$

Figure 6.8 shows $x_L(\lambda)$, $x_H(\lambda)$, and $K(\lambda) = x_H(\lambda)/x_L(\lambda)$ for the case $Q_L = Q_H = 1$, and $q_L = 10$, $q_H = 1$,

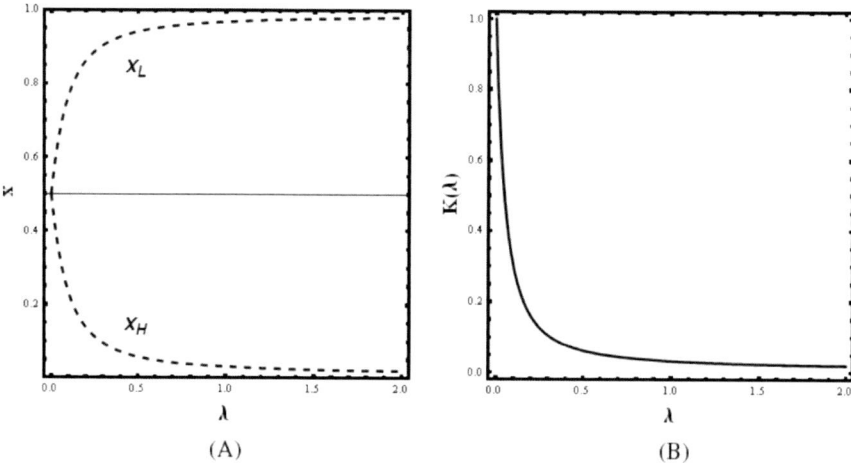

Fig. 6.8 (A) $X_L(\lambda)$ and $X_H(\lambda)$ as given by Eq. (6.47). (B) $K(\lambda) = X_H(\lambda)/X_L(\lambda)$.

$S_L = S_H = S = 1$. Clearly, since $q_L/q_H > 1$, x_L increases with λ. The values of $x_L^{(0)} x_L^{(1)}$ and $x_L^{(2)}$ are in this case 0.5, 0.909 and 0.99099, respectively. The value of $x_L = 0.99099$ is the limiting value of x_L at $\lambda \to \infty$, or $\theta \to 1$.

Two other useful probabilities are

$$Pr(1, 1) = (Q_L q_L^2 S_L + Q_H q_H^2 S_H)\lambda^2/\xi$$
$$Pr(1) = Pr(1, \_) = Pr(\_, 1) = (Q_L q_L + Q_H q_H)\lambda/\xi$$
$$(6.51)$$

The first is the probability of finding an adsorbent molecule doubly occupied — the conformational state of the molecule is unspecified. Clearly, this is the sum of two probabilities

$$Pr(1, 1) = Pr(L; 1, 1) + Pr(H; 1, 1) \qquad (6.52)$$

The pair correlation function is defined by

$$g(1, 1) = \frac{Pr(1, 1)}{Pr(1)^2} \qquad (6.53)$$

As we discussed in Section 6.1, all we need is the $\lambda \to 0$ limit of this correlation function which is

$$g^0(1, 1) = \lim_{\lambda \to 0} [g(1, 1)] = \frac{Q(1, 1)Q(0, 0)}{Q(1, 0)Q(0, 1)} \qquad (6.54)$$

For the remainder of this book we will refer to $g^0(1, 1)$ as the pair correlation, and drop the superscript zero.

In terms of the intrinsic binding constants we have

$$k_{11} = g(1, 1)k_1^2 \qquad (6.55)$$

Using the notation $K = Q_H/Q_L$ and $h = q_H/q_L$, we rewrite Eq. (6.54) in the form

$$g(1, 1) = \frac{(S_L + S_H Kh^2)(1 + K)}{(1 + Kh)^2} \qquad (6.56)$$

We see that the correlation function $g(1, 1)$ is not simply related to the *direct* correlations S_L and S_H. Clearly, this is not an average of the two *direct* correlations. In this section we wish to focus on the *indirect* correlation. Therefore, we assume that the direct correlations are either negligible, i.e. $S_L \approx S_L \approx 1$, or that they are independent of the conformation, i.e. $S_L = S_L = S$. Hence, $g(1, 1)$ may be written as

$$g(1, 1) = y(1, 1)S \qquad (6.57)$$

The factor $y(1, 1)$ as defined in (6.57) is referred to as the *indirect* correlation between the two ligands. It can be written in the form:

$$y(1,1) = \frac{(1+Kh^2)(1+K)}{(1+Kh)^2} = 1 + \frac{K(1-h)^2}{(1+Kh)^2} \quad (6.58)$$

The indirect correlation is the major source of cooperativity in biochemical systems, such as hemoglobin or allosteric enzymes. The model treated in this section is the simplest binding model having indirect correlation. We will now examine some of the outstanding properties of the indirect correlation $y(1,1)$.

First, note that $y(1,1) = 1$ (i.e. no indirect correlation) if, and only if, either $K = 0$, or $h = 1$. This follows directly from the equality on the rhs of Eq. (6.58). Incidentally, the condition for the occurrence of an indirect correlation is exactly the same as the necessary and sufficient condition for inducing conformation changes by the ligand. Therefore, it follows that indirect correlation occurs if, and only if, the ligand induces conformational changes in the adsorbent molecules.[3]

For any fixed value of h, $y(1,1)$ has a maximum as a function of K at the point $K = h^{-1}$. Figure 6.9 shows $y(1,1)$ as a function of K for several values of h. At the point $K = h^{-1}$, the maximal value of $y(1,1)$ is

$$y_{max} = \frac{2+h+h^{-1}}{4} = \frac{2+K+K^{-1}}{4} \quad (6.59)$$

Recall that $K = \exp[-\beta(E_H - E_L)]$ and $h = \exp[-\beta(U_H - U_L)]$. Therefore, for very large h (or very small $K = h^{-1}$), y_{max} is determined either by the difference of

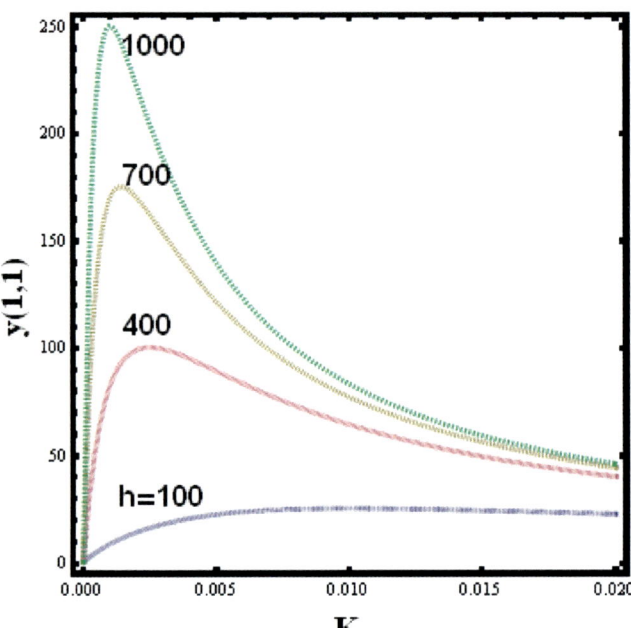

Fig. 6.9 $y(1, 1)$ as a function of K for several values of h, as indicated.

the binding energies $U_H - U_L$, or by the difference of the energies $E_H - E_L$.

It is clear from (6.58) that the indirect correlation is always positive, i.e. $y(1, 1) \geq 1$. We also see from Eq. (6.58) and from Eq. (6.59) that y_{max} can be very large, depending on the values of $E_H - E_L$, or $U_H - U_L$, but is independent of S or $U(1, 1)$. Thus, whenever conformational changes are induced by the binding process, we should find a *positive* contribution to the cooperativity. The physical reason for this behavior is quite simple. When the ligand has a preference for one conformation, i.e. $h \neq 1$, the binding of the first ligand will shift the equilibrium $L \rightleftharpoons H$ toward the species that it favors. The second ligand approaching the second site will find a new equilibrium concentration of L and H, namely,

$x_L^{(1)}$ and $x_H^{(1)}$. Hence, the conditional probability of binding to the second site is larger than the (unconditional) probability of binding to the first site.

In terms of the free energies we have $W(1, 1) = -k_B T \ln[g(1, 1)]$ and, assuming $S_L = S_H = S$, we have the equality

$$W(1, 1) = U(1, 1)$$
$$+ [A(1, 1) + A(0, 0) - A(0, 1) - A(1, 0)]$$
$$= U(1, 1) + \delta A \qquad (6.60)$$

δA being the indirect part of the work associated with the process

$$(0, 1) + (1, 0) \rightarrow (1, 1) + (0, 0) \qquad (6.61)$$

Exercise: Show that the quantity δA, defined in (6.60) is always negative, i.e. $\delta A \leq 0$.

The quantity δA can also be rewritten as

$$\delta A = [A(1, 1) - A(0, 1) - \mu] - [A(1, 0) - A(0, 0) - \mu]$$
$$= \Delta A^*(1|1) - \Delta A^*(1|0) < 0 \qquad (6.62)$$

where $\Delta A^*(1|0)$ and $\Delta A^*(1|1)$ are the binding free energies on the first and second sites, respectively.

We now summarize the main difference between the *direct* and the *indirect* correlations:

1. The sign of the direct correlation depends on the *direct* interaction between the ligands. The sign of the indirect correlation is *always* positive (for two identical sites; see below for two different sites), and it is independent of

$U(1, 1)$, but dependent on the difference of binding energies $U_H - U_L$, and the difference of energies of the two conformers $E_H - E_L$.

2. The direct interaction depends on the distance between the ligands and has the same range as the ligand-ligand pair potential. The indirect correlation, in this particular model, is *independent* of the ligand-ligand distance. It depends on the capacity of the ligand to induce conformational changes ($h \neq 1$), and on the responsiveness of the adsorbing molecule ($K \neq 0$).

3. $U(1, 1)$ is presumed to be temperature-independent, while $W(1, 1)$ is in general temperature-dependent. In terms of correlations, the sign of the temperature dependence of S depends on the sign of $U(1, 1)$. If $U(1, 1) < 0$, then S decreases with T; when $U(1, 1) > 0$, S increases with T. On the other hand, the dependence of $y(1, 1)$ on T is determined by the enthalpy change associated with the process (6.61). In biochemical processes the *indirect* correlation can cause an enormous increase in the cooperativity, far beyond the normally weak, or even negligible, *direct* cooperativity which depends on the ligand-ligand interactions (say, two oxygen molecules in hemoglobin). This spectacular and sophisticated trick has been selected by evolution searching for more efficient ways of regulating biochemical processes.[4]

The temperature dependence of $y(1, 1)$ is determined by

$$\frac{\partial \ln y}{\partial T} = \frac{-\Delta H^*}{k_B T^2} \sim \frac{-\Delta E^*}{k_B T^2} \qquad (6.63)$$

where ΔH^* and ΔE^* are the enthalpy and the energy associated with the indirect part of the work $W(1, 1)$, i.e.

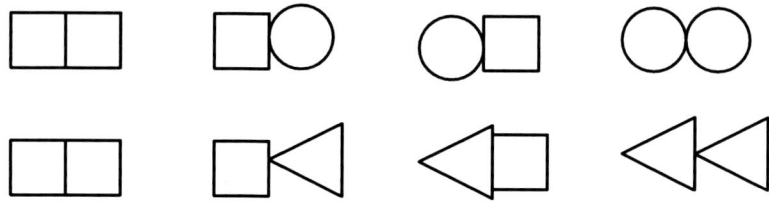

Fig. 6.10 Two subunits with symmetric (lower panel) and asymmetric (upper panel) interactions.

with δA in Eq. (6.60). It is easy to show that $|S|$ is always a monotonically decreasing function of T, while y can increase or decrease with T for a given set of molecular parameters, h and K. This fact could serve as a diagnostic test for the existence of an indirect correlation.

We have started our discussion of the properties of $y(1, 1)$ with the assumption $S_L = S_H = S$. This led to the factorization of $g(1, 1)$ as in Eq. (6.57). It is easy to see that all the properties of $y(1, 1)$ hold true also when $S_L \neq S_H$. In this case, Eq. (6.56) can be rewritten as

$$g(1, 1) = \frac{(1 + Kh^2)(1 + K)}{(1 + Kh)^2} \frac{(S_L + S_H Kh^2)}{(1 + Kh^2)}$$
$$= y(1, 1)\langle S \rangle \tag{6.64}$$

Here, $y(1, 1)$ is exactly the same as in Eq. (6.58). The factor S in Eq. (6.57) is now replaced by an average of S_L and S_H, with weights $(1 + Kh^2)^{-1}$ and $Kh^2/(1 + Kh^2)$, respectively.

We now briefly discuss the case of two *different* sites for which the PF is

$$\xi = (Q_L + Q_H) + (Q_L q_{La} + Q_H q_{Ha})\lambda$$

$$+ (Q_L q_{Lb} + Q_H q_{Hb})\lambda$$
$$+ (Q_L q_{La} q_{Lb} + Q_H q_{Ha} q_{Hb})S\lambda^2 \qquad (6.65)$$

where we have used the notation

$$q_{La} = \exp(-\beta U_{La}), \quad q_{Lb} = \exp(-\beta U_{Lb})$$
$$q_{Ha} = \exp(-\beta U_{Ha}), \quad q_{Hb} = \exp(-\beta U_{Hb}) \qquad (6.66)$$

The binding isotherm for this case is

$$\bar{n} = \frac{(k_a + k_b)C + 2k_a k_b g_{ab} C^2}{1 + (k_a + k_b)C + k_a k_b g_{ab} C^2} \qquad (6.67)$$

where g_{ab} is defined by

$$g_{ab} = g(a, b) = \frac{Q(a, b)Q(0, 0)}{Q(a, 0)Q(0, b)} \qquad (6.68)$$

Denoting

$$K = \frac{Q_H}{Q_L}, \quad h_a = \frac{q_{Ha}}{q_{La}}, \quad h_b = \frac{q_{Hb}}{q_{Lb}} \qquad (6.69)$$

and assuming that $S_L = S_H = S$, we can write $y(a, b)$ in the form

$$y(a, b) = \frac{(1 + h_a h_b K)(1 + K)}{(1 + h_a K)(1 + h_b K)}$$
$$= 1 + \frac{(1 - h_a)(1 - h_b)K}{(1 + h_a K)(1 + h_b K)} \qquad (6.70)$$

The important difference between this case and the previous one in Eq. (6.58) is that here the indirect cooperativity can be either positive or negative, depending on whether the signs of $(1 - h_a)$ and $(1 - h_b)$ are the same or different.

Thus, if $h_a > 1$ and $h_b < 1$, binding on a will shift the equilibrium $L \rightleftharpoons H$ toward H (the favored conformation when the ligands bind to site a), but since $h_b < 1$, this means that the new conformational equilibrium will be less favorable for binding on site b. In this case

$$P(b|a) < P(b) \tag{6.71}$$

or, equivalently

$$A(a, b) + A(0, 0) - A(a, 0) - A(0, b) > 0$$

or

$$\Delta A^*(b|a) - \Delta A^*(b) > 0 \tag{6.72}$$

6.5 Indirect Correlation Due to Transmission of Information Across the Boundary Between the Subunits

We next extend the model of Section 6.4 by one aspect. The adsorbent molecules now consist of two identical subunits, each having one binding site. The subunit itself can be in one of two conformations, L or H. Hence, altogether there are *four* possible states for the entire empty adsorbent molecule: LL, LH, HL and HH. Formally, this model extends the model discussed in Section 6.4 and allows for four states instead of two. In this respect all the results of the previous model also apply to this model, and in some special cases ($\eta \to 0$, see below) the two models actually become identical.

The study of this model is important because many real biological systems consist of subunits. The resulting mechanism of cooperativity is prevalent in biochemical systems.

6.5.1 *The Empty System*

The system consists of two subunits, each of which can attain one of two states, denoted by L and H, having energies E_L and E_H, respectively. In addition, we have inter-subunit interactions, which we denote by E_{LL}, $E_{LH} = E_{HL}$, and E_{HH} depending on the state of the two subunits. Note that in general, E_{LH} can differ from E_{HL}. However, for simplicity we assume that $E_{LH} = E_{HL}$, as shown in Fig. 6.10. Denote

$$Q_\alpha = \exp(-\beta E_\alpha), \quad Q_{\alpha\beta} = \exp(-\beta E_{\alpha\beta}) \quad (6.73)$$

where subscripts α and β can be either L or H.

Let us first examine the equilibrium properties of this system in the absence of ligands. The grand PF of a single system is

$$\xi(0) = Q_L^2 Q_{LL} + 2Q_L Q_H Q_{LH} + Q_H^2 Q_{HH} \quad (6.74)$$

Note that this is actually the canonical partition function for a single empty system. It is also the limit of the grand PF of the system for $\lambda \to 0$.

The four states of the empty system are LL, LH, HL and HH (see the first row of Fig. 6.11) with corresponding probabilities, or mole fractions,

$$x_{LL}^0 = \frac{Q_L^2 Q_{LL}}{\xi(0)}, \quad x_{LH}^0 = x_{HL}^0 = \frac{Q_L Q_H Q_{LH}}{\xi(0)},$$

$$x_{HH}^0 = \frac{Q_H^2 Q_{HH}}{\xi(0)} \quad (6.75)$$

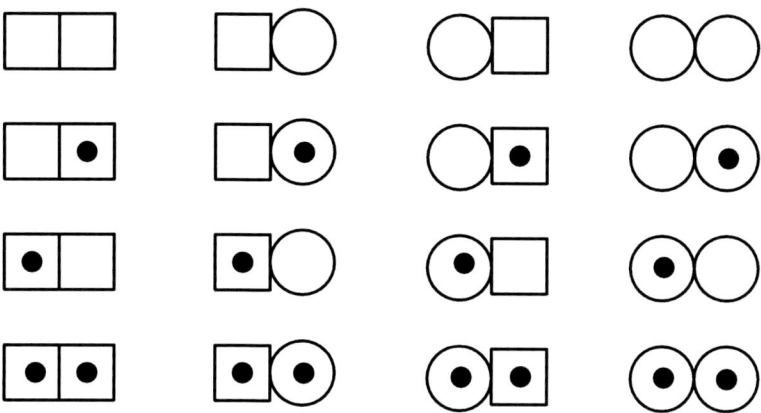

Fig. 6.11 All sixteen configurations of a system with two subunits.

In Eq. (6.75) the superscript zero indicates an empty system, i.e. the absence of ligands. Note that x_{LH}^0 is the probability of finding, let's say, the rhs subunits in the H state and the lhs subunits in the L state. The mole fraction of the system is such that any one of the subunits is in the L state, and the other in the H state and is the sum of x_{LH}^0 and x_{HL}^0, which in our case is simply $2x_{LH}^0$.

We next define the probability of finding a specific subunit (say the lhs) in state L, independently of the state of the second subunit, by

$$x_L^0 = x_{LL}^0 + x_{LH}^0 = \frac{Q_L^2 Q_{LL} + Q_L Q_H Q_{LH}}{\xi(0)} \tag{6.76}$$

and, similarly

$$x_H^0 = x_{HH}^0 + x_{HL}^0 = \frac{Q_H^2 Q_{HH} + Q_L Q_H Q_{LH}}{\xi(0)} \tag{6.77}$$

It follows from the definition (6.75) that in the limit of infinite separation between the two subunits, there will be no interactions between the subunits, hence we have

$$x_{\alpha\beta}^{0,\infty} \rightarrow \frac{Q_\alpha Q_\beta}{(Q_L + Q_H)^2} = x_\alpha^{0,\infty} x_\beta^{0,\infty} \tag{6.78}$$

which is the expected result; i.e. the probability of finding one subunit in state α, and the second in state β is simply the product of the two probabilities. We conclude that $E_{\alpha\beta} = 0$ (for $\alpha, \beta = L, H$) is a *sufficient* condition for independence of the subunits. This is not a necessary condition, however.

There is another important case where the two subunits become independent. This occurs when the interaction energies between the subunits are such that

$$E_{LH} = \frac{1}{2}(E_{LL} + E_{HH}) \tag{6.79}$$

or, equivalently when

$$Q_{LH}^2 = Q_{LL} Q_{HH} \tag{6.80}$$

In this case the PF can be written as

$$\xi(0) = Q_L^2 Q_{LL} + 2Q_L Q_H \sqrt{Q_{LL} Q_{HH}} + Q_H^2 Q_{HH}$$
$$= (Q_L \sqrt{Q_{LL}} + Q_H \sqrt{Q_{HH}})^2 \tag{6.81}$$

From Eqs. (6.75) and (6.76), we find

$$x_L^0 x_L^0 = \frac{(Q_L^2 Q_{LL} + Q_L Q_H \sqrt{Q_{LL} Q_{HH}})^2}{(Q_L \sqrt{Q_{LL}} + Q_H \sqrt{Q_{HH}})^4}$$

$$= \frac{Q_L^2 Q_{LL}}{(Q_L \sqrt{Q_{LL}} + Q_H \sqrt{Q_{HH}})^2} = x_{LL}^0 \tag{6.82}$$

and, similarly

$$x_{LL}^0 = x_L^0 x_L^0, \quad x_{LH}^0 = x_L^0 x_H^0, \quad x_{HH}^0 = x_H^0 x_H^0 \tag{6.83}$$

Thus, whenever the condition (6.79) or (6.80) is fulfilled, the probability of the states of the two subunits is the product of the probabilities of the states of each subunit. We see that the existence of interaction between the subunits does not imply dependence of the states of the subunits. In other words, the condition $E_{\alpha\beta} = 0$ is not a necessary condition for independence.

Consider the reaction

$$(LL) + (HH) \rightarrow 2(LH) \tag{6.84}$$

For this reaction, the equilibrium constant is

$$\eta = \frac{(x_{LH}^0)^2}{x_{LL}^0 x_{HH}^0} = \frac{(Q_L Q_H Q_{LH})^2}{Q_L^2 Q_{LL} Q_H^2 Q_{HH}} = \frac{Q_{LH}^2}{Q_{LL} Q_{HH}}$$
$$= \exp[-\beta(2E_{LH} - E_{LL} - E_{HH})] \tag{6.85}$$

Thus, in the empty system the equilibrium constant η is determined only by the interaction energies $E_{\alpha\beta}$. The condition (6.79) is equivalent to the condition $\eta = 1$. We will later see that the equilibrium constant η is also responsible for the transmission of information between the two ligands across the boundary between the two subunits.

It is easy to show that if Eq. (6.83) holds, i.e. if $x_{\alpha\beta}^0$ is a product of x_α^0 and x_β^0 as defined in Eqs. (6.76) and (6.77), then $\eta = 1$. This follows directly from the definitions of $x_{\alpha\beta}^0$, x_α^0 and x_β^0. Therefore, $\eta = 1$ is a *necessary* and *sufficient* condition for independence of the *states* of the two subunits. We will see that this statement holds true also for the *ligands* occupying the sites of the system.

6.5.2 *The Binding Isotherm*

We next introduce ligands into the system. The binding energies to states L and H are U_L and U_H, respectively, and we assume also direct interaction energy $U(1, 1)$ between the ligands (for simplicity. We further assume that this is independent of the states of the subunits). We denote this by

$$q_L = \exp(-\beta U_L), \quad q_H = \exp(-\beta U_H),$$
$$S = \exp[-\beta U(1, 1)] \tag{6.86}$$

The grand PF for the system is now

$$\xi = Q(0; 0) + [Q(0; 1) + Q(1; 0)]\lambda + Q(1; 1)\lambda^2$$
$$= \sum_{\alpha\beta} Q_\alpha Q_\beta Q_{\alpha\beta}$$

$$+ \left(\sum_{\alpha\beta} Q_\alpha Q_\beta q_\alpha Q_{\alpha\beta} + \sum_{\alpha\beta} Q_\alpha Q_\beta q_\beta Q_{\alpha\beta} \right) \lambda$$

$$+ \sum_{\alpha\beta} Q_\alpha Q_\beta Q_{\alpha\beta} q_\alpha q_\beta S\lambda^2 \tag{6.87}$$

The sixteen terms in this equation correspond to the sixteen states of the system in Fig. 6.11. Each sum in Eq. (6.87) corresponds to one row in Fig. 6.11. We use the semicolon to separate the specification of the states of the left and of the right subunits. Note also that we assume for simplicity, that $S_L = S_H = S$.

Another way of writing the PF in Eq. (6.87) is

$$\xi = \sum_{\alpha\beta} Q_\alpha Q_\beta Q_{\alpha\beta}[1 + (q_\alpha + q_\beta)\lambda + q_\alpha q_\beta S\lambda^2] \tag{6.88}$$

Here, each term with specific α and β, e.g. $\alpha = L$ and $\beta = L$ correspond to one conformational state of the system, i.e. one *column* in Fig. 6.11. For instance,

$$\xi_{LL} = Q_L^2 Q_{LL}(1 + 2q_L\lambda + q_L^2 S\lambda^2) \tag{6.89}$$

is the PF for the system in the conformational state LL. Clearly, this is the same as the PF of a single-state system with two binding sites. The general form of the binding isotherm for this model is

$$\theta = \frac{\bar{n}}{2} = \frac{\lambda}{2}\frac{\partial \ln \xi}{\partial \lambda} = \frac{\sum_{\alpha\beta} Q_\alpha Q_\beta Q_{\alpha\beta}(q_\alpha\lambda + q_\alpha q_\beta S\lambda^2)}{\xi} \tag{6.90}$$

Note that even when $S = 1$, the binding isotherm does not reduce to the Langmuir form. If, on the other hand, both $S = 1$, and $\eta = 1$, then we find that the PF is

$$\xi = \sum_{\alpha\beta} Q_\alpha Q_\beta \sqrt{Q_{\alpha\alpha} Q_{\beta\beta}}(1 + q_\alpha\lambda)(1 + q_\beta\lambda)$$

$$= \left[\sum_\alpha Q_\alpha \sqrt{Q_{\alpha\alpha}}(1 + q_\alpha\lambda)\right]^2 \tag{6.91}$$

Which is essentially the same PF as that of the model treated in Section 6.4 with $S_L = S_H = 1$. We will see in the following subsections that the important new features of this model arise when $\eta \neq 1$.

As in Section 6.4 we define the intrinsic binding constant by

$$k_1 = \frac{Q(0; 1)}{Q(0; 0)}\lambda_0 \tag{6.92}$$

and the correlation function

$$g(1, 1) = \frac{Q(1; 1)Q(0; 0)}{Q(0; 1)Q(1; 0)} \tag{6.93}$$

We rewrite the binding isotherm as

$$\theta = \frac{k_1 C + k_1^2 g(1, 1)C^2}{1 + 2k_1 C + k_1^2 g(1, 1)C^2} \tag{6.94}$$

which is formally the same as in Section 6.2, but with k_1 replacing k and $g(1, 1)$ replacing S.

6.5.3 Correlation Function and Cooperativity

The correlation function was defined in Section 6.1 for any two events. In particular, if the events are "site one is occupied" and "site two is occupied," then the (λ-dependent) correlation function is

$$g(1, 1) = \frac{Pr(1; 1)}{Pr(1)^2} = \frac{Q(1; 1)\lambda^2 \xi}{[Q(0; 1)\lambda + Q(1; 1)\lambda^2]^2} \tag{6.95}$$

Note here that $Pr(1; 1)$ is the probability of finding the two sites occupied (independently of the states of the subunits) and $Pr(1)$ is the probability of finding a *specific* site (say, the lhs in Fig. 6.11 occupied (independently of the state of the subunits and of the occupational state of the second site); $Pr(1) = Pr(1; \_) = Pr(\_; 1)$.

As in Section 6.4, we showed that we need only the $\lambda \rightarrow 0$ limit of this correlation function, which is the quantity defined in Eq. (6.93), which we refer to as the correlation function between the two events. For these particular *events* we also say that whenever there exists correlation [i.e. $g(1, 1) \neq 1$], the two *ligands cooperate*; hence, there exists *cooperativity* between the ligands, or simply the system is cooperative.

We now focus on the *indirect* correlation. To do so, we can either factor out S from $g(1, 1)$, or simply assume that $S = 1$, and examine the remaining correlation, denoted by $y(1, 1)$. Using the notations (6.85) and (6.86), and $\bar{K} = Q_H/Q_L$, and $K' = Q_{HH}/Q_{LL}$, we note that $\bar{K}$ and $\sqrt{K'}$ always appear together. Hence, we define

$$K = \bar{K}\sqrt{K'} = \frac{Q_H\sqrt{Q_{HH}}}{Q_L\sqrt{Q_{LL}}} \qquad (6.96)$$

and write the indirect correlation in the simplified form

$$y(1, 1) = \frac{(1 + 2\bar{\eta}K + K^2)(1 + 2\bar{\eta}hK + h^2K^2)}{(1 + \bar{\eta}K + \bar{\eta}hK + hK^2)^2}$$

$$= 1 - \frac{(\bar{\eta}^2 - 1)(h - 1)^2K^2}{(1 + \bar{\eta}K + \bar{\eta}hK + hK^2)^2} \qquad (6.97)$$

The second form on the rhs of this equation is very convenient for examining the condition under which indirect correlation exists. In Eq. (6.97) and in the following section we put $\bar{\eta} = \sqrt{\eta}$.

We can compare Eq. (6.97) with Eq. (6.58) in Section 6.4. We will find that the former reduces to the latter when $\eta = 0$, and $K^2 = K$. In the model of Section 6.4, we found that "either $h = 1$, or $K = 0$," is a necessary and sufficient condition for $y(1, 1) = 1$. In the present model, we see from Eq. (6.97) that "either $h = 1$, or $K = 0$, or $\eta = 1$," is a necessary and sufficient condition for $y(1, 1) = 1$.

As in the model of Section 6.4, if $K = 0$ (or $K = \infty$), the system is not responsive to the binding process. Also, if $h = 1$, the ligand cannot induce conformational changes in the system. Here, in addition to the requirement $K \neq 0$, and $h \neq 1$, we need $\eta \neq 1$ in order to get an indirect correlation. Moreover, the correlation $y(1, 1)$ in Section 6.4 is always positive $[y(1, 1) \geq 1]$. Here, the correlation may be either positive or negative, depending on whether $\eta < 1$ or $\eta > 1$. It can be shown that this property is related to the way the conformation change induced in one subunit is transmitted to the second subunit.[5] Clearly, if all $E_{\alpha\beta} = 0$, then there can be no communication between the two subunits, hence $y(1, 1) = 1$. However, this result is also obtained under much weaker conditions, namely when $2E_{LH} = E_{LL} + E_{HH}$, or equivalently, $\eta = 1$. Thus, when E_{LH} equals the average of E_{LL} and E_{HH}, i.e. $\eta = 1$, then $y(1, 1) = 1$. The reverse is also true as can be seen from Eq. (6.97).

To summarize, we see that in order to have indirect cooperativity the following conditions must be fulfilled.

(a) The ligand must have a preference for one of the conformational states, i.e. it must be able to *induce* conformational change. The capacity to do so is measured by h. When $h = 1$, no conformational change can be induced.

(b) The subunit must be responsive. If $K = 0$ (or $K = \infty$, but we have chosen $0 \leq K \leq 1$), the conformational state will not respond to the binding process. The system remains in its most stable form (which was chosen to be the L form). To be responsive, the system must have a non-zero equilibrium constant, K.

(c) Whatever the conformational change induced in the subunit on which the ligand is bound, there must be another conformational change induced in the second subunit. The extent and the sign of this change is measured by η. If $\eta < 1$, the change induced in the second subunit will be in the *same direction* as in the first; hence the approaching second ligand will find a subunit, the state of which is the preferable state for binding, hence $y(1, 1) > 1$. The reverse is true when $\eta > 1$, i.e. negative cooperativity. When $\eta = 1$, there is no transmission of information across the boundaries, hence, no indirect cooperativity. This is a very general scheme of the conditions that must be fulfilled to obtain indirect correlation.[5]

6.5.4 *Exercises*

We discuss here very briefly two important limiting cases. The details are left as exercises.[5]

1. The concerted model

The first limiting case was suggested originally by Monod, Wyman and Changeux (MWC), Monod *et al.* (1965). This model requires that the two subunits be either in the L or in the H state. The conformations of the two subunits change in a *concerted* way. This is equivalent to the consideration of the first and fourth columns in Fig. 6.11. Mathematically, we can obtain this limiting case by taking $\eta = 0$, which essentially means that the equilibrium concentrations of X_{LH} (and X_{HL}) are negligible.

The PF of the system may be obtained from Eq. (6.88) by substituting $Q_{\alpha\beta}^2 = Q_{\alpha\alpha} Q_{\beta\beta} \delta_{\alpha\beta}$. The result is

$$\xi_{MWC} = \sum_\alpha Q_\alpha^2 Q_{\alpha\alpha} (1 + 2 q_\alpha \lambda + q_\alpha^2 S \lambda^2) = \xi_{LL} + \xi_{HH}$$

(6.98)

This is essentially the same PF as that of the model treated in Section 6.4 with the replacement of Q_α everywhere by $Q_\alpha^2 Q_{\alpha\alpha}$. Thus, the MWC model is equivalent to a two-state model with energies corresponding to $2E_L + E_{LL}$ and $2E_H + E_{HH}$. The fact that we have two subunits does not affect the formalism, except for the redefinitions of the energy levels.

Originally, the MWC was applied to systems with negligible direct interactions between the ligands (i.e. $S = 1$), in which case Eq. (6.98) is reduced to

$$\xi_{MWC} = Q_L^2 Q_{LL} (1 + q_L \lambda)^2 + Q_H^2 Q_{HH} (1 + q_H \lambda)^2$$

(6.99)

If we choose the L energy level, $2E_L + E_{LL}$, as our zero energy (or equivalently, define a new PF by $\xi' = \xi / Q_L^2 Q_{LL}$), we can rewrite Eq. (6.99) in a more familiar form

$$\xi'_{MWC} = (1 + K_L C)^2 + K (1 + K_H C)^2 \qquad (6.100)$$

where $K_L C = q_L \lambda$, $K_H C = q_H \lambda$ and K is the equilibrium constant for conversion between the two states, $LL \rightleftharpoons HH$:

$$K = \frac{Q_H^2 Q_{HH}}{Q_L^2 Q_{LL}} \qquad (6.101)$$

From Eq. (6.97), we see that in the case $\eta = 0$, the indirect correlation function (which is the same as the total correlation function if $S = 1$) is

$$y(1, 1) = 1 + \frac{(1 - h)^2 K}{(1 - Kh)^2} \geq 1 \qquad (6.102)$$

which is essentially the same as in Eq. (6.58) with the replacement of K in (6.58) by K in (6.101). In this case, as in the model of Section 6.4, we always have positive (indirect) cooperativity. The molecular reason for this result is quite clear in view of the analysis of the origin of the cooperativity as discussed in Section 6.4. Here, a conformational change on one subunit is *fully transmitted* to the second subunit. In other words, the two subunits respond in a *concerted* manner, as if they were a single subunit, i.e. as in the model of Section 6.4.

2. The sequential model

The second extreme case, suggested by Koshland, Nemethy and Filmer (KNF), Koshland *et al.* (1996), is also known as the *sequential* mode. The mathematical conditions required to obtain this limiting case are quite severe. First, it is assumed that in the absence of a ligand, one of the conformations is dominant, say the LL form. In addition, it is assumed that a ligand binding to any subunit will change the conformation of that subunit into the H form. These assumptions lead to the consideration of only four diagonal states of Fig. 6.11, for which the PF is

$$\xi_{KNF} = Q_L^2 Q_{LL} + 2 Q_L Q_H Q_{LH} q_H \lambda + Q_H^2 Q_{HH} q_H^2 S \lambda^2 \qquad (6.103)$$

The empty state is the LL state on the top left corner of Fig. 6.11. The binding of a ligand on any of the subunits will shift its conformation completely from L to H without affecting the conformation of the second subunit. Binding of the two ligands will shift the entire system to the state HH. Thus, in each binding process there is a *total* change of conformation of one subunit; hence, the term *sequential* model.

The mathematical requirements necessary to obtain the KNF model from the general one can be stated as follows. Let $K^{(0)}$ be the equilibrium constant for the $H \rightleftharpoons L$ conversion of each subunit when it is known to be empty. Likewise, let $K^{(1)}$ be the equilibrium constant when the subunit is known to be occupied. Both of these equilibrium constants are functions of λ, i.e.

$$K^{(0)} = \frac{Pr(H, 0;)}{Pr(L, 0;)} = \frac{Q_H}{Q_L} F(\lambda) \qquad (6.104)$$

and

$$K^{(1)} = \frac{Pr(H, 1;)}{Pr(L, 1;)} = \frac{Q_H q_H}{Q_L q_L} F(S\lambda) \qquad (6.105)$$

where $Pr(H, 0;)$ is the probability of finding the left-hand subunits in state H and being empty. Similar meanings apply to the other probabilities in Eqs. (6.104) and (6.105). The functions $F(\lambda)$ and $F(S\lambda)$ may be obtained from the PF, through the relevant probabilities.

The mathematical requirement of the model is that $K = Q_H/Q_L \rightarrow 0$, i.e. L is infinitely more stable than H, but $Kh = Q_H q_L/Q_L q_L \rightarrow \infty$, i.e. the occupied subunit in *state* H is infinitely more stable than L. This means that we require a *total conversion* $L \rightarrow H$, upon binding in a system which has *infinite resistance* ($K \to 0$) to conformational changes. This can be achieved only when h is such a "strong infinity" that even after multiplication by $K \to 0$, the product Kh is still infinity. Clearly, this model is not realistic. However, it was an important model in the history of studying cooperative phenomena.

6.5.5 *Connection between Theory and Experiments*

Perhaps, the simplest two-site cooperative systems are small molecules having two binding sites for protons, such as dicarboxylic acids and diamines. Despite their molecular simplicity, the theoretical treatment of their behavior is much more intricate. The main reasons for this are: (1) there is, in general, a continuous range of macrostates; (2) the direct and indirect correlations are both strong and intertwined, so that factorization of the correlation function is impossible. In addition, the solvent might have a major effect on the binding properties of these molecules.

The literature on dissociation (or ionization) constants of these and related compounds is immense. In this section we will present only a few examples to illustrate the kind of cooperativity in these systems.

We will use only *intrinsic binding* (or *association*) constants. The relevant experimental data are usually reported in terms of either the *thermodynamic* dissociation or association constant. The general relationship between the intrinsic and the thermodynamic constants is discussed below.

We denote by k_1 the first intrinsic binding constant, and by k_2 the *second* intrinsic binding constant. The latter is

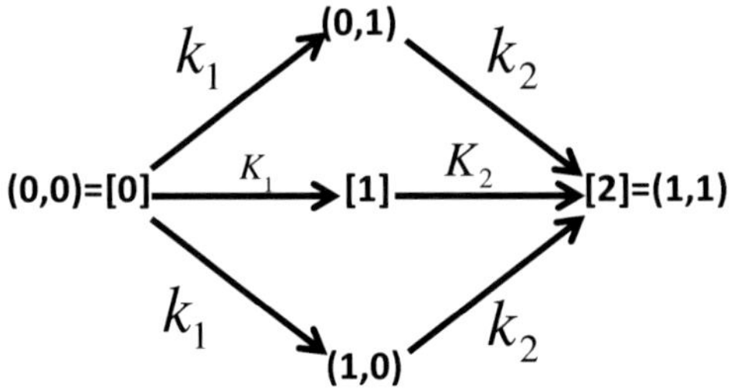

Fig. 6.12 Two alternative ways of describing the binding of protons to a dicarboxylic acid. [0], [1] and [2] denote the di-ion, singly and doubly occupied acid, respectively. $(0, 1)$ and $(1, 0)$ indicate a molecule with a specific single site occupied. The *macroscopic* binding constants K_1 and K_2 are related to the reactions of adding successively the first and second proton to the di-ion, respectively. The *intrinsic* binding constants k_1 and k_2 refer to the same processes but for a *specific* site, as if the two (identical) binding sites were labeled, say the right and the left one.

the conditional binding on the second site, given that the first site is occupied. Figure 6.12 shows two alternatives, but equivalent ways of describing the binding of protons to a dicarboxylic acid.

(a) The first view referred to as the "macroscopic" view, recognizes *three* species: the empty, the singly occupied, and the doubly occupied molecules. The corresponding densities of the species are denoted by square brackets. The two ("macroscopic") equilibrium constants are

$$K_1 = \frac{[1]}{[0][H]}, \quad K_2 = \frac{[2]}{[1][H]} \tag{6.106}$$

where $[H]$ is the proton density (in moles per unit of volume) in the solution.

(b) The second view, sometimes referred to as the "micro-scopic" view, recognizes *four* species denoted by $(0, 0)$,

$(0, 1)$, $(1, 0)$ and $(1, 1)$. Here, we "label" the two equivalent "sites," say the right and left sides, and distinguish between configurations $(0, 1)$ and $(1, 0)$. These configurations are in general indistinguishable (unless the molecules are localized, or if we consider two non-equivalent sites, say in salicylic acid, or in amino acids).

The "microscopic" equilibrium constants are defined by

$$k_1 = \frac{[0, 1]}{[0, 0][H]} = \frac{[1, 0]}{[0, 0][H]} \qquad (6.107)$$

$$k_2 = \frac{[1, 1]}{[0, 1][H]} = \frac{[1, 1]}{[1, 0][H]} \qquad (6.108)$$

Note that the terms "macroscopic" and "microscopic" constants do not imply that these quantities measure macroscopic or microscopic quantities, respectively. Here, in the "macroscopic" view we have simply grouped the two "microscopic" species $(0, 1)$ and $(1, 0)$ into one species denoted by (1). Both these constants can be macroscopic or microscopic, depending on whether we study the binding per molecule or per one mole of molecules.

Since we have the condition

$$[1] = [0, 1] + [1, 0] \qquad (6.109)$$

The "macroscopic" and "microscopic" constants are related by:

$$K_1 = 2k_1 \quad \text{and} \quad K_2 = k_2/2 \qquad (6.110)$$

The factor 2 in the first relationship arises from having two "products" in the first association process (Fig. 6.12). The factor $1/2$ in the second arises from having two "reactants" from which the doubly occupied acid is formed. (These factors are sometimes referred to as "statistical"; a better term would be combinatorial factors). The binding isotherm (or the equivalent titration curve) should not depend on the way we choose to view our species. If we denote by C the proton density $[H]$, in the solution (assuming that we are in the regime of very dilute solutions), then the binding isotherm is simply the ratio between the total bound protons and the total number of adsorbent molecules, i.e.

$$\bar{n} = \frac{[0,1] + [1,0] + 2[1,1]}{[0,0] + [0,1] + [1,0] + [1,1]}$$
$$= \frac{K_1 C + 2 K_1 K_2 C^2}{1 + K_1 C + K_1 K_2 C^2}$$
$$= \frac{2 k_1 C + 2 k_1 k_2 C^2}{1 + 2 k_1 C + k_1 k_2 C^2} \tag{6.111}$$

The distinction between K_i and k_i becomes clearer when the two sites are not equivalent (e.g. glycerine), in which case $2k_1$ in Eq. (6.111) is the sum of two *different* intrinsic constants, say the binding to the acidic group k_a and to the basic group, k_b such as

$$2 k_1 = k_a + k_b \tag{6.112}$$

From the experimental binding isotherm $\bar{n} = \bar{n}(C)$, one can determine the two "experimental" constants K_1 and K_2. If the two sites are identical, then one can convert from K_1 and K_2 into k_1 and k_2 via Eq. (6.110). This is not possible when the two sites are not identical in which case one cannot obtain the intrinsic constants, say k_a and k_b, solely from knowledge of K_1 and K_2.

The thermodynamic dissociation constants are related to the association constants by

$$K_1 = (K_{2,diss})^{-1} \quad \text{and} \quad K_2 = (K_{1,diss})^{-1} \qquad (6.113)$$

The pair correlation function for the two protons is given by

$$g_{11} = g(1, 1) = \frac{k_2}{k_1} = \frac{4k_2}{K_1} = \frac{4K_{2,diss}}{K_{1,diss}} \qquad (6.114)$$

The corresponding work is defined by

$$W_{11} = W(1, 1) = -k_B T \ln g_{11} \qquad (6.115)$$

This is the work associated with the process

$$(0, 1) + (1, 0) \rightarrow (1, 1) + (0, 0) \qquad (6.116)$$

In this process we start from two singly occupied molecules, but at infinite separation from each other (and hence no correlation between the protons), and form one doubly occupied and one empty molecule. By taking *two* singly occupied *molecules* at infinite separation, we secure the independence of the protons at the same two sites as in the doubly occupied molecule, denoted $(1, 1)$. The quantity W_{11} is computed from the pK values, usually reported in the literature as $T = 298.15K$ (here, K stands for degrees Kelvin).

$$W_{11} = -k_B T \ln \frac{4K_2}{K_1} = -k_B T 2.303 \log_{10} \frac{4K_2}{K_1}$$
$$= 1.364(pK_2 - pK_1 - \log_{10} 4) \qquad (6.117)$$

Where $pK = -\log K$ is the logarithm to base 10 of the binding constant K and $2.303 k_B T = 1.364 \, kcal/mol$ is the value for conversion to $kcal/mol$ at $T = 298.15 \, K$.

In the next subsections we will present some specific experimental data. In all cases known to us, W_{11} is positive, indicating negative cooperativity. Most molecular interpretations of this cooperativity have focused on the electrostatic interaction between the two protons. Formally, as shown below, one can always define a *microscopic* dielectric constant to account for the deviation in the experimental value of W_{11} from the value computed on the basis of purely Coulombic interaction. However, from the theoretical point of view, we can gain further insight into the molecular contributions to W_{11} by distinguishing between the following four sources of cooperativity; (1) *direct* proton–proton interaction; (2) *indirect* proton–proton correlation mediated by the solvent; (3) indirect correlation mediated by the adsorbent molecule at a fixed conformation; (4) indirect correlation, modified by the allowance of conformational changes in the adsorbent molecule.

Qualitatively, the four components of the correlation function can be visualized according to the following four steps: First, we bring the two protons from infinite separation to the final distance R_{HH} (where R_{HH} is assumed to be a *fixed* distance between the two protons on one of the conformational states of the adsorbent molecule). The interaction energy is simply the Coulombic interaction e^2/R_{HH}, where e is the charge of the proton. Next, we place the same two protons in a solvent. The interaction is now modified by the dielectric constant of the solvent. If the distance R_{HH} is large enough, one can use the macroscopic dielectric constant of pure water, which at 25°C is 78.5. However, for smaller distances, the dielectric constant would depend on the distance R_{HH} (and, in general also on the size and type of the charges on the two interacting species). Next, having selected a fixed conformation, we put the adsorbent molecule between the two protons. This will modify the correlation between the two protons. Finally, a further modification will be obtained by relaxing the requirement of a fixed conformation.

Perhaps, the first attempt to interpret the proton–proton correlation, based on electrostatic interactions was made by Bjerrum (1923). If e is the proton charge and R_{HH} the proton–proton distance in the diacid, then the interaction free energy was assumed to be given by

$$W_{11} = \frac{e^2}{DR_{HH}} \qquad (6.118)$$

where D is the macroscopic dielectric constant (78.5 for water). One can fit the experimental values of W_{11} to an equation of the form (6.118) only for long α, ω-dicarboxylic acids where the protons are far apart as will be illustrated in the next subsection (see Fig. 6.13). For the shorter diacids, W_{11} is much larger than the value predicted with the fixed dielectric constant D.

Clearly, if one takes a smaller value of D, one gets a larger value of W_{11} for a given distance R_{HH}. In fact one can argue that one should take a much smaller dielectric constant, since the medium between the two protons more closely resembles a hydrocarbon liquid rather than water. In fact, for any dicarboxylic acid one can *define* an effective dielectric constant D_E to fit the experimental value of W_{11} by an equation of the form (6.118), with D_E being dependent

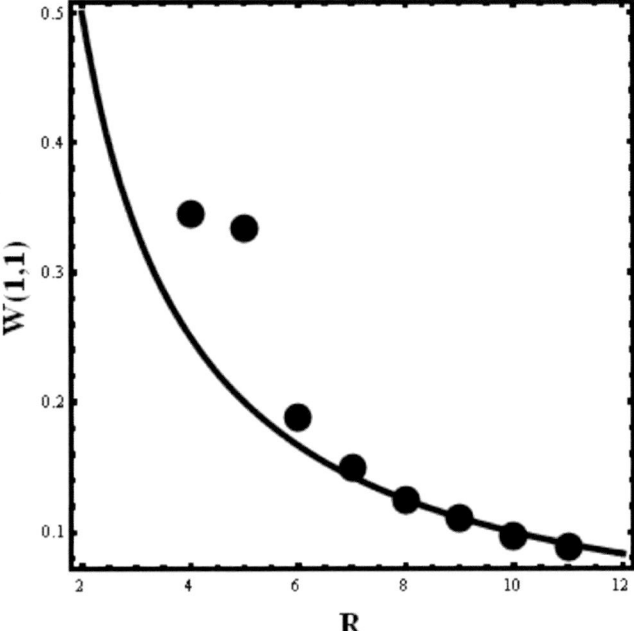

Fig. 6.13 Experimental values of $W(1, 1)$ on α, ω-dicarboxylic acids. The proton–proton distances were calculated for the fully extended linear acids. The full curve is $W(1, 1) = e^2/78.5 \, R_{HH}$ in kcal/mol.

on the proton–proton distance, the type and size of the acid and the solvent.

In 1959, Eberson (1959, 1992) found that a family of derivatives of succinic acid shows a remarkably large negative cooperativity, i.e. $g_{11} \ll 1$, which is difficult to explain on the basis of the electrostatic theories alone. At present, there is no satisfactory molecular interpretation of these findings. One of the more popular ideas is that an intramolecular hydrogen bond would facilitate the first dissociation of the proton, i.e. $K_{1,diss}$ becomes smaller (or K_2 becomes larger). Also, the second proton will dissociate with more difficulty. The net effect would be a decrease in the correlation function

Table 6.1 First and second intrinsic binding constants (in liter/mol), pair correlations and corresponding work (1,1) (in kcal/mol) for α, ω-dicarboxylic acid $COOH(CH_2)_n COOH$ at 25°C.

Acid	k_1	k_2	$g(1, 1)$	$W(1, 1)$
Oxalic acid ($n = 0$)	7.8×10^4	3.39×10^1	4.34×10^{-3}	3.22
Malonic acid ($n = 1$)	2.46×10^5	1.34×10^4	5.45×10^{-3}	3.09
Succinic acid ($n = 2$)	1.50×10^5	3.12×10^4	2.08×10^{-1}	0.93
Glutaric acid ($n = 3$)	1.31×10^5	4.41×10^4	3.35×10^{-1}	0.65
Adipic acid ($n = 4$)	1.29×10^5	5.23×10^4	4.05×10^{-1}	0.53
Pimelic acid ($n = 5$)	1.33×10^5	6.10×10^4	4.60×10^{-1}	0.46
Suberic acid ($n = 6$)	1.26×10^5	6.51×10^4	5.15×10^{-1}	0.39
Azelaic acid ($n = 7$)	1.30×10^5	7.12×10^4	5.48×10^{-1}	0.36

g_{11}. There are other factors that can possibly contribute to the correlation function, one being specific solvent effects such as formation of a hydrogen-bond bridge by solvent molecules.[6]

Two identical sites: dicarboxylic acids and diamines

Table 6.1 shows some values of k_1, k_2, the correlation function $g(1, 1)$, and the corresponding free energy $W(1, 1)$

for a series of α, ω–dicarboxylic acids. We first note that all the correlations $g(1, 1) < 1$, i.e. negative cooperativity. As expected, the cooperativity is strongest for oxalic acid for which the proton–proton distance is about 3 Å; the value of $g(1, 1)$ gradually increases toward 1 for the larger molecules. For the largest dicarboxylic acids reported here, the azelaic acid, where seven methylene groups separate the two carboxylic groups, we still find substantial correlation between the two protons. The largest distance is about 10.6 Å. Note that in the limit of very large n, the *second* binding constant should approach the value of k for the monocarboxylic acid, $k \approx 7.2 \times 10^4$; the value of k_2 for azelaic acid is 7.12×10^4.

In Fig. 6.13, we plot the experimental free energies $W(1, 1)$ as a function of the proton–proton distance R_{HH}. We also plot the theoretical curve for the Coulombic interaction between the two protons, as modified by the *macroscopic* dielectric constant of water $D = 78.54$. It is clear that the values $W(1, 1)$ for the larger molecules follow closely the theoretical curve with a fixed value of D. Large deviations occur for the smaller molecule. This may be interpreted as due to a much smaller effective dielectric constant.

Table 6.2 presents some values of k_1, k_2, $g(1, 1)$ and $W(1, 1)$ for some α, ω–alkane diamines. Again, we see that as the proton–proton distance (in the di-ionized molecule) increases, the value of $g(1, 1)$ increases toward unity, but even for the 1,8 octane diamine the correlation is still quite large (0.504). In contrast to the case of dicarboxylic acid, the α, ω–diamines cannot be fitted on a Coulombic curve of the form (6.118) with the macroscopic dielectric constant of water.

Table 6.2 First and second intrinsic binding constants (in liter/mol), pair correlations, and corresponding work ($W(1,1)$) (in kcal/mol) for α, ω-alkane diamines at 20°C.

Diamine	k_1	k_2	$g(1,1)$	$W(1,1)$
1,2 Ethane diamine ($n = 2$)	6.15×10^9	2.00×10^7	3.25×10^{-3}	3.34
1,3 Propane diamine ($n = 3$)	2.08×10^{10}	8.73×10^8	4.19×10^{-2}	1.85
1,4 Butane diamine ($n = 4$)	3.15×10^{10}	4.48×10^9	1.42×10^{-1}	1.14
1,5 Pentane diamine ($n = 5$)	5.61×10^{10}	1.10×10^{10}	1.96×10^{-1}	0.95
1,8 Octane diamine ($n = 8$)	5.00×10^{10}	2.52×10^{10}	5.04×10^{-1}	0.40

The best one can do is to fit these data to a Coulombic curve with an effective dielectric constant of $D_E \approx 28$. As in the case of dicarboxylic acids, we find here that for the long chain diamine the first binding constant is about 5×10^{10}, very close to the limit of the k values of the normal mono-amines, which is about 4.4×10^{10}.

Maleic, fumaric and succinic acids

Table 6.3 shows some values of the proton–proton correlation for the three dicarboxylic acids: maleic, fumaric and succinic acids (Fig. 6.14). All the values of $W(1,1)$ are positive, i.e. negative cooperativity. Since the configuration of the first two acids is nearly rigid, one can expect that the larger the proton–proton distance, the weaker the cooperativity.

Table 6.3 Experimental values of $g(1, 1)$ and $w(1, 1)$ (in kcal/mol) for maleic, fumaric, and succinic acid.

	In water at 25		In 50% aqueous ethanol at 20°	
	$g(1, 1)$	$W(1, 1)$	$g(1, 1)$	$W(1, 1)$
Maleic	1.52×10^{-4}	5.21	1.0×10^{-6}	8.18
Fumaric	1.75×10^{-1}	1.034	6.05×10^{-2}	1.66
Succinic	2.05×10^{-1}	0.938	4.92×10^{-2}	1.78

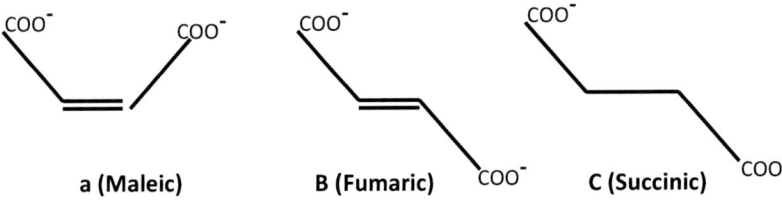

a (Maleic)   B (Fumaric)   C (Succinic)

Fig. 6.14 The structural formulae for maleic, fumaric and succinic acids.

Indeed, the ratio of $W(1, 1)$ for the first two acids is

$$\frac{W(maleic)}{W(fumaric)} = \begin{cases} 5.04 & (for\ water\ at\ 25°C) \\ 4.93 & (for\ 50\%\ aqueous\ ethanol\ at\ 20°C) \end{cases}$$

$$(6.119)$$

Note that the ratio is almost the same for the two solvents. These ratios are, however far from the inverse ratio of the proton–proton distances, which is about

$$\frac{R_{HH}(fumaric)}{R_{HH}(maleic)} \approx 1.94 \qquad (6.120)$$

Thus, one cannot fit a function of the form

$$W(1, 1) = \frac{e^2}{DR_{HH}} \qquad (6.121)$$

for the two isomers with the same dielectric constant.

The case of the succinic acid cannot be discussed in terms of the Coulombic interactions alone. Here, conformational changes induced by the binding process can contribute significantly to the correlation. Note also that $g(1, 1)$, [or $W(1, 1)$] of succinic acid is not an "average" of the correlations in maleic and fumaric acids. This could be partially due to the configurational changes in the succinic acid, induced by the binding process.[7]

7

One-Dimensional Lattice Models

One-dimensional (1D) models are extremely useful in studying a myriad of molecular phenomena. Before you skip reading this chapter, being convinced that the system you are studying is not a 1D system — hold on! Your system does not need to be a 1D system. You can learn a great deal about your system even if in reality your system "lives" in a three-dimensional space.

There are several reasons why I believe you might benefit from studying the theory of 1D systems. First, there are some actual systems which are linear, and for which a 1D model might be a good approximate description. Second, there are many systems that "live" in three-dimensional space but whose properties can be studied in any dimension. In such cases it is advantageous to study that system in 1D.

For these reasons, I urge you not to skip this chapter just because your problem is not a 1D problem. Besides, the study of thermodynamics of 1D systems is simple, elegant and beautiful.

Once you are proficient in solving 1D problems, you will have a powerful tool, which can be applied to obtain answers to a variety of problems. We have plenty of examples in the rest of this book where a phenomena featuring in a 3D system can be studied and understood by a simple solvable 1D model.

In this chapter we will study the simplest 1D models; particles occupying lattice points in 1D space. The emphasis here is on the technique of solving for the PF of such a system. We will study three apparently different systems. However, as we will soon see, all these systems are formally equivalent, but we use the different systems to study different aspects of the response of such systems to an external "field". For instance, in the system of magnetic spins we can define the average mole fractions of particles in the "up" and "down" states. These mole fractions can be changed by an external magnetic field. Likewise, the mole fractions of sites that are in the two conformations L and H can be changed by an external pressure (or tension), or by the ligand concentration.

7.1 A System of Magnetic Dipoles

We consider a linear system of M sites on which there are $N = M$ molecules, each having a spin. The spins can be in one of two states, up or down. We associate with each spin a variable $s_i (i = 1, 2, \ldots, N)$, which is $+1$ if the ith spin is in the state "up" and -1 if its state is "down". The configuration of the entire system is described by the N dimensional vector $s = (s_1, s_2, \ldots, s_N)$. For instance a possible configuration of a system of six spins is described by the vector $s = (1, -1, 1, 1, -1, 1)$. See Fig. 7.1.

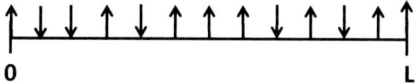

Fig. 7.1 A 1D system of spins "up" and "down".

Each spin interacts only with its nearest neighbors. The interaction energy between ith and $(i + 1)th$ is given by

$$E_{i,i+1} = -\varepsilon s_i s_{i+1} \tag{7.1}$$

We assume that $\varepsilon > 0$. When two spins align in the same direction, they contribute a negative interaction energy $-\varepsilon$. When they are aligned in opposite directions they contribute positive energy. The case $\varepsilon > 0$ corresponds to ferromagnetism and the case $\varepsilon < 0$ to anti-ferromagnetism.

In addition to the assumption of pairwise interactions (7.1), each spin interacts with an external magnetic field H, which is constant throughout the system. The energy of the system in any given configuration $(s_1, \ldots, s_N)$ is:

$$E(\mathbf{s}) = -\varepsilon \sum_{i=1}^{N} s_i s_{i+1} - H \sum_{i=1}^{N} s_i \tag{7.2}$$

H is chosen in such a way that for $s_i = +1$, the interaction with the magnetic field is negative; i.e. the state "up" is energetically more favorable in the presence of a magnetic field.

The canonical partition function of the system of N spins at temperature T and subjected to an external magnetic field H is given by

$$Q(N, H, T) = \sum_{s} \exp[-\beta E(\mathbf{s})] \tag{7.3}$$

where the summation is over all possible states or configurations of the system, i.e. all the possible vectors $\mathbf{s} = (s_1, s_2, \ldots, s_N)$ of the system. Since each spin can attain one of the two states $s_1 = \pm 1$ the total number of configurations is 2^N. We now rewrite the sum (7.3) in more detailed form as

$$Q(N, H, T) = \sum_{s_1 = \pm 1} \sum_{s_2 = \pm 1} \cdots \sum_{s_N = \pm 1}$$
$$\times \exp[(\beta H s_1 + \varepsilon s_1 s_2 + H s_2 + \varepsilon s_2 s_3$$
$$+ \cdots H s_{N-1} + \varepsilon s_{N-1} s_N + H s_N)] \quad (7.4)$$

We further assume that N is very large so that edge (surface) effects are negligible. Hence, we can close the system into a circle; i.e. the first particle, $i = 1$ interacts with the last particle $i = N$. See Fig. 7.2.

The closing of the system requires the addition of the factor $\exp(\beta \varepsilon s_N s_1)$ to the sum (7.4). It is easy to see that such an addition has no significant effect on the thermodynamics of the system provided that N is very large.

Take note that each term in the sum (7.4) has alternating factors: One factor of $\exp(\beta H s_i)$ for each *particle*, and one factor of $\exp(\beta \varepsilon s_i s_{i+1})$ for each *bond*, i.e. a pair of consecutive particles. Altogether we have $2N$ factors. It is convenient to rewrite the terms in (7.4) as a product of N factors. There are several ways of doing so. One is to assign to each *bond*

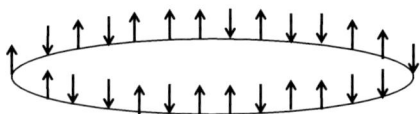

Fig. 7.2 Closing the 1D system into a cycle.

the factor

$$\exp{(\beta H s_i/2)} \exp{(\beta \varepsilon s_i s_{i+1})} \exp{(\beta H s_{i+1}/2)} \qquad (7.5)$$

In this way, each term in (7.4) consists of N factors, each pertaining to one bond.[1]

The factor corresponding to the $i, i+1$ bond is now denoted by

$$P_{i,i+1} = \langle s_i | P | s_{i+1} \rangle$$
$$= \exp{(\beta H s_i/2)} \exp{(\beta \varepsilon s_i s_{i+1})} \exp{(\beta H s_{i+1}/2)}$$
$$(7.6)$$

Since there are two states for each particle $s_i = \pm 1$, there are altogether four possible *bonds*. The corresponding factors for these four bonds are

$$\langle +1|P|+1 \rangle = \exp[\beta(\varepsilon + H)]$$
$$\langle +1|P|-1 \rangle = \exp[\beta(-\varepsilon)]$$
$$\langle -1|P|+1 \rangle = \exp[\beta(-\varepsilon)]$$
$$\langle -1|P|-1 \rangle = \exp[\beta(\varepsilon - H)] \qquad (7.7)$$

These four factors define a 2×2 matrix that has the form

$$P = \begin{pmatrix} \exp[\beta(\varepsilon + H)] & \exp{(-\beta\varepsilon)} \\ \exp{(-\beta\varepsilon)} & \exp[\beta(\varepsilon - H)] \end{pmatrix} \qquad (7.8)$$

We also define the row-vectors

$$\langle +1| = (1, 0), \quad \langle -1| = (0, 1) \qquad (7.9)$$

and the column-vectors

$$|+1\rangle = \begin{pmatrix} 1 \\ 0 \end{pmatrix}, \quad |-1\rangle = \begin{pmatrix} 0 \\ 1 \end{pmatrix} \qquad (7.10)$$

With these vectors we can generate the four terms in (7.7) by matrix multiplications:

$$\langle +1|P|+1 \rangle = (1,0)P\begin{pmatrix}1\\0\end{pmatrix} = \exp[\beta(\varepsilon + H)]$$

$$\langle +1|P|-1 \rangle = (1,0)P\begin{pmatrix}0\\1\end{pmatrix} = \exp[-\beta\varepsilon]$$

$$\langle -1|P|+1 \rangle = (0,1)P\begin{pmatrix}1\\0\end{pmatrix} = \exp[-\beta\varepsilon]$$

$$\langle -1|P|-1 \rangle = (0,1)P\begin{pmatrix}0\\1\end{pmatrix} = \exp[\beta(\varepsilon - H)] \quad (7.11)$$

We now write the PF in (7.4) in matrix notation as follows

$$Q(N, H, T) = \sum_s \exp\left(\beta H s_1 / 2\right)\langle s_1|P|s_2 \rangle \cdots$$

$$\times \langle s_{N-1}|P|s_N \rangle \exp[\beta H s_N / 2] \quad (7.12)$$

We now use the multiplication rule for two $N \times N$ matrices A and B. The element ij of the product AB is

$$(AB)_{ij} = \sum_{k=1}^N A_{ik} B_{kj} \quad (7.13)$$

where the sum is over the index k.

Applying this rule to all the indices $s_2 s_3 \cdots s_{N-1}$ gives the element $\langle s_1|P^{N-1}|s_N \rangle$. Hence, (7.12) is rewritten as

$$Q(N, H, T)$$
$$= \sum_{s_1 s_N} \exp\left(\beta H s_1 / 2\right)\langle s_1|P^{N-1}|s_N \rangle \exp\left(\beta H s_N / 2\right)$$

$$(7.14)$$

We now add the factor $\exp{(\beta \varepsilon s_N s_1)}$ to close the cycle. We can simplify (7.14) by summing over s_1 and s_N to obtain the final form

$$Q(N, H, T) = \sum_{s_1 s_N} \langle s_1 | P^{N-1} | s_N \rangle \langle s_N | P | s_1 \rangle$$

$$= \sum_{s_1} \langle s_1 | P^N | s_1 \rangle = Tr P^N \qquad (7.15)$$

Thus, closing the circle allows us to write the PF of the system as a trace of the matrix P^N. This leads to an elegant solution of the one-dimensional Ising model. The trace of matrix A is defined as the sum of all its diagonal elements, i.e.

$$Tr A = \sum_i A_{ii}$$

Once we have the diagonal elements of a 2×2 matrix we can solve for the PF in terms of the molecular parameters of the model. To achieve this we use a theorem from matrix algebra that states that if A is a matrix that diagonalizes the matrix P, it will also diagonalize the matrix P^N.

Let A be the matrix that diagonalizes P, i.e.

$$APA^{-1} = \vec{\lambda} \qquad (7.16)$$

where λ is a diagonal matrix, i.e.

$$\lambda = \begin{pmatrix} \lambda_1 & 0 \\ 0 & \lambda_2 \end{pmatrix} \qquad (7.17)$$

λ_1 and λ_2 are the eigenvalues of the matrix A, and A^{-1} is the inverse of A. It is easy to show that the matrix A also diagonalizes P^N, i.e.

$$
\begin{aligned}
AP^N A^{-1} &= APPP \cdots PA^{-1} \\
&= APA^{-1}APA^{-1}APA^{-1} \cdots APA^{-1} \\
&= \lambda^N = \begin{pmatrix} \lambda_1^N & 0 \\ 0 & \lambda_2^N \end{pmatrix}
\end{aligned}
\tag{7.18}
$$

In (7.18) we have introduced the unit matrix $A^{-1}A = I$ between each pair of Ps.

It is also easy to show that the trace of any product of matrices is invariant to cyclic permutation of matrices, i.e.

$$
Tr(AB) = Tr(BA) \tag{7.19}
$$

This follows directly from the definition of a trace:

$$
Tr(AB) = \sum_i \sum_k A_{ik} B_{ki} = \sum_k \sum_i B_{ki} A_{ik} = Tr(BA)
\tag{7.20}
$$

Therefore, it follows that

$$
Tr(APA^{-1}) = Tr(A^{-1}AP) = Tr(P) \tag{7.21}
$$

Applying (7.21) to the PF (7.15), we obtain the final result

$$
Q(N, H, T) = TrP^N = Tr(AP^N A^{-1}) = \lambda_1^N + \lambda_2^N
\tag{7.22}
$$

Thus, the calculation of the PF reduces to the calculation of the *eigenvalues* of the 2×2 matrix P. Since we will be interested only in the limit of large N, we will need only the larger of the two *eigenvalues*. Suppose that $\lambda_1 > \lambda_2$. (In the case of $\lambda_1 = \lambda_2$ we have a degeneracy, and this should be discussed separately.[2] The Helmholtz energy of the system is

$$A = -k_B T \ln Q = -k_B T \ln (\lambda_1^N + \lambda_2^N)$$
$$= -k_B T \ln \lambda_1^N - k_B T \ln[1 + (\lambda_2/\lambda_1)^N] \quad (7.23)$$

When $N \to \infty$ the second term on the rhs of (7.23) can be neglected. Hence, the PF depends only on the largest eigenvalue of the matrix P.

Our task now is to calculate the largest eigenvalue in terms of the molecular parameters of the system. The eigenvalues of the matrix P may be obtained from the solution of the secular equation, which is

$$|P - \lambda I| = 0 \quad (7.24)$$

where I is the 2×2 matrix. More specifically,

$$|P - \lambda I| = \begin{vmatrix} \exp[\beta(\varepsilon + H)] - \lambda & \exp[-\beta\varepsilon] \\ \exp[-\beta\varepsilon] & \exp[\beta(\varepsilon - H)] - \lambda \end{vmatrix}$$
$$= 0 \quad (7.25)$$

or equivalently

$$\{\exp[\beta(\varepsilon + H)] - \lambda\}\{\exp[\beta(\varepsilon - H)] - \lambda\}$$
$$- \exp(-2\beta\varepsilon) = 0 \quad (7.26)$$

This is a polynomial in λ, the two roots of which are

$$\lambda_\pm = \exp(\beta\varepsilon)\left[\cosh x \pm \sqrt{\sinh^2 x + \exp(-4\beta\varepsilon)}\right] \quad (7.27)$$

where $x = \beta H$, and we used the identities

$$\sinh x = \frac{e^x - e^{-x}}{2}, \quad \cosh x = \frac{e^x + e^{-x}}{2},$$

$$\cosh^2 x - \sinh^2 x = 1 \tag{7.28}$$

It is easy to check that the two roots are positive, and that $\lambda_+ > \lambda_-$. Thus, we have an explicit expression for the Helmholtz energy in terms of the molecular parameters of the model:

$$A = -k_B T \ln Q = -k_B T \ln \lambda_+^N \tag{7.29}$$

The most important quantity of our system is the magnetization M_a defined by

$$M_a = k_B T \left(\frac{\partial \ln Q}{\partial H} \right)_{T,N}$$

$$= k_B T \left[\frac{\sum_s \exp[-\beta E(s) \sum_i \beta s_i]}{Q} \right]$$

$$= \sum_s P(s) \sum_i s_i = \left\langle \sum_{i=1}^N s_i \right\rangle = N \langle s_i \rangle \tag{7.30}$$

The meaning of M_a is quite simple. For any specific configuration s, we denote the number of spins in the "up" state by N_+ and the number of spins in the "down" state by N_-. Therefore, for any specific configuration s, we have

$$M_a(s) = \sum_{i=1}^N s_i = N_+ - N_- \tag{7.31}$$

Thus, M_a in (7.30) is the average of $N_+ - N_-$. (The magnetization is actually this number times the magnetic moment m_0).

From the PF (7.29) we obtain for the magnetization of the system

$$M_a(N, H, T) = k_B T \frac{\partial \ln \lambda_+}{\partial H}$$

$$= N \frac{\sinh(\beta H)}{[\sinh^2(\beta H) + \exp(-\beta 4\varepsilon)]^{1/2}}$$

$$(7.32)$$

If there are no interactions $\varepsilon = 0$, the spins are independent and

$$M_a = N \frac{\sinh(\beta H)}{\cosh(\beta H)} = N \tanh(\beta H) \qquad (7.33)$$

Figure 7.3 shows the magnetization per particle as a function of βH for different values of $\beta \varepsilon$. When H is positive most of the spins will be in the same direction as H, i.e. in the "up" state. The direction is changed to "down" when H is negative.

We see that for $\beta \varepsilon = 0$ the magnetization per particle changes continuously from all-down to all-up as βH changes from negative to positive values. What is more interesting is when we "turn-on" the interactions ($\beta \varepsilon > 0$), the curves change more sharply from -1 to $+1$. The reason is that the interaction between the spins "cooperates" in the sense that if a spin is in the "up" state, it "encourages" the neighboring spin to align in the same direction. This is true for the case of ferromagnetism ($\varepsilon > 0$). Note also that the larger the value of ε, the sharper the transition from negative to positive

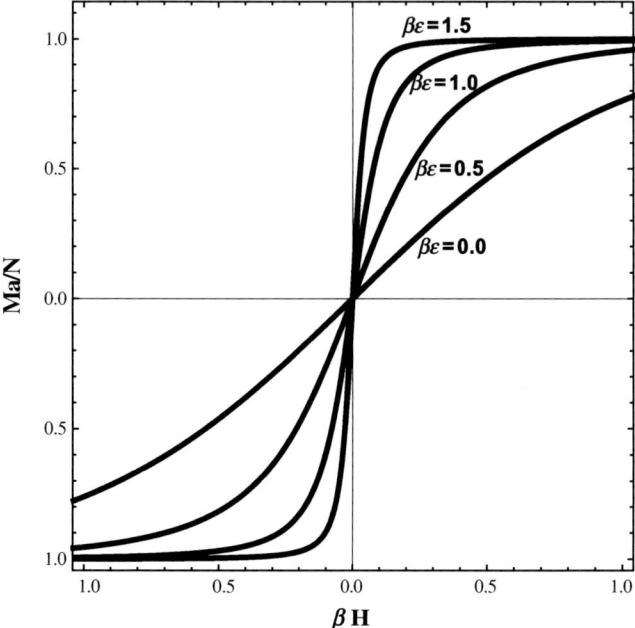

Fig. 7.3 The magnetization per particle M_a/N as a function of βH for different values of $\beta \varepsilon$ (as indicated next to each curve).

Fig. 7.4 A lattice model for adsorption.

magnetization. For very large values of ε the curve resembles a phase transition curve.

7.2 Adsorption on a Lattice with Nearest Neighbors Interactions

This is a simple variation of the Ising model discussed in Section 7.1. Again, we have M cells that we refer to as the lattice points. Each lattice point can either be empty or occupied by a particle, see Fig. 7.4. The canonical PF of a system of N particles on M lattice points $(N < M)$ is:

$$Q(T, M, N) = \sum_{\mathbf{s}} \exp[-\beta E(\mathbf{s})] \qquad (7.34)$$

where again the configuration of the entire system is described by the vector $\mathbf{s} = (s_1, \ldots, s_M)$, with $s_i = 0$ and $s_i = 1$ for empty and occupied sites, respectively.

Let W_{11} be the interaction energy between two particles on adjacent sites. In this model $W_{01} = W_{00} = 0$; i.e. there is no interaction between a particle and an empty site or between two empty sites. Let N_{11} be the number of pairs of nearest neighbors particles at any given configuration s. Clearly, the energy levels of the system are determined by N_{11}, i.e.

$$E(\mathbf{s}) = N_{11} W_{11} \qquad (7.35)$$

The partition function (7.34) is written as

$$Q(T, M, N) = \sum_{\substack{energy \\ levels}} \Omega(N_{11}) \exp\left(-\beta N_{11} W_{11}\right)$$

$$= \sum_{\substack{all\ s\ with \\ fixed\ N}} \exp\left(-\beta N_{11} W_{11}\right) \qquad (7.36)$$

The first sum on the rhs of (7.36) is over all energy levels. Since the energy levels are determined by the parameter N_{11}, this is the same as summation over all possible values of N_{11}. $\Omega(N_{11})$ is the number of configurations (states of the system) with a fixed N_{11}, or equivalently, with a fixed energy level $N_{11} W_{11}$. The second sum is over all possible configurations of the system.

In this particular example it is possible to compute $\Omega(N_{11})$ and evaluate the canonical partition function of the system. A more elegant solution is obtained for the grand PF using the same matrix method as was used in previous sections. Thus, we open the system with respect to the particles and write the grand PF as

$$
\Xi(T, M, \lambda) = \sum_{N=0}^{M} Q(T, M, N)\lambda^{N}
$$

$$
= \sum_{N=0}^{M} \lambda^{N} \sum_{\substack{all\ s\ with \\ fixed\ N}} \exp\left(-\beta N_{11} W_{11}\right)
$$

$$
= \sum_{All\ s} \lambda^{N} \exp\left(-\beta N_{11} W_{11}\right) \qquad (7.37)
$$

where λ is the absolute activity of the adsorbed particles, or the ligands.

Note that each term in the last sum on the rhs of (7.37) corresponds to one specific vector **s**. For each configuration there are N factors λ corresponding to the N particles, and a factor $q = \exp[-\beta W_{11}]$ for each particle–particle interaction, see Fig. 7.5.

We assigned a factor 1 for a pair $(0, 0)$, a factor $\lambda^{1/2}$ for a pair $(0, 1)$, or a pair $(1, 0)$ and a factor λq for a pair $(1, 1)$.

1	1	0	0	1	1	0	0
$\lambda^{1/2}q\lambda^{1/2}$	$\lambda^{1/2}$	1	$\lambda^{1/2}$	$\lambda^{1/2}q\lambda^{1/2}$	$\lambda^{1/2}$	1

Fig. 7.5 Assigning factors to the different neighbors.

We define the matrix elements

$$\langle 0|P|0\rangle = 1$$
$$\langle 0|P|1\rangle = \langle 1|P|0\rangle = \lambda^{1/2}$$
$$\langle 1|P|1\rangle = \lambda q \tag{7.38}$$

As in the previous section we define the row-vectors

$$\langle 0| = (0, 1), \quad \langle 1| = (1, 0) \tag{7.39}$$

and the corresponding column-vectors

$$|0\rangle = \begin{pmatrix} 0 \\ 1 \end{pmatrix}, \quad |1\rangle = \begin{pmatrix} 1 \\ 0 \end{pmatrix} \tag{7.40}$$

Thus, the elements of the matrix P in (7.39) may be obtained by multiplication of row vector, the matrix P and a column-vector.

The grand PF (7.37) may be written after closing the cycle (i.e. the Mth site becomes nearest neighbor to first site 1.)

$$\Xi(T, M, \lambda)$$
$$= \sum_s \langle s_1|P|s_2\rangle \langle s_2|P|s_3\rangle \cdots \langle s_{M-1}|P|s_M\rangle \langle s_M|P|s_1\rangle$$
$$= Tr P^M \tag{7.41}$$

where the matrix P is defined by its elements in (7.39), i.e.

$$P = \begin{pmatrix} 1 & \lambda^{1/2} \\ \lambda^{1/2} & \lambda q \end{pmatrix} \tag{7.42}$$

In order to compute the trace of P^M, we need the eigenvalues of the matrix P. The mathematical problem is the same as in Section 7.1. We denote by γ the eigenvalues of P. Thus, we need to solve the secular question

$$|P - \gamma I| = 0 \tag{7.43}$$

or equivalently,

$$\begin{vmatrix} 1 - \gamma & \lambda^{1/2} \\ \lambda^{1/2} & \lambda q - \gamma \end{vmatrix} = 0 \tag{7.44}$$

The two solutions of (7.44) are

$$\gamma_- = \frac{(1 + \lambda q) - \sqrt{(1 + \lambda q)^2 + 4\lambda}}{2} \tag{7.45}$$

$$\gamma_+ = \frac{(1 + \lambda q) + \sqrt{(1 + \lambda q)^2 + 4\lambda}}{2} \tag{7.46}$$

Clearly, $\gamma_+ > \gamma_-$. Therefore, for $M \to \infty$ we need only γ_+. Hence, in this limit the grand PF is

$$\Xi(T, M, \lambda) = \gamma_+^M \tag{7.47}$$

The average number of particles in the system is

$$\bar{N} = \lambda \frac{\partial \ln \Xi}{\partial \lambda} = M \frac{\lambda[qr - q(1 - \lambda q) + 2]}{(1 + \lambda q)r + 1(1 - \lambda q)^2 + 4\lambda} \tag{7.48}$$

where $r = \sqrt{(1 - \lambda q)^2 + 4\lambda}$.

Equation (7.47) can also be written in a more convenient way as

$$\theta = \frac{\bar{N}}{M} = \frac{1 - \gamma_+}{1 + \lambda q - 2\gamma_+} \tag{7.49}$$

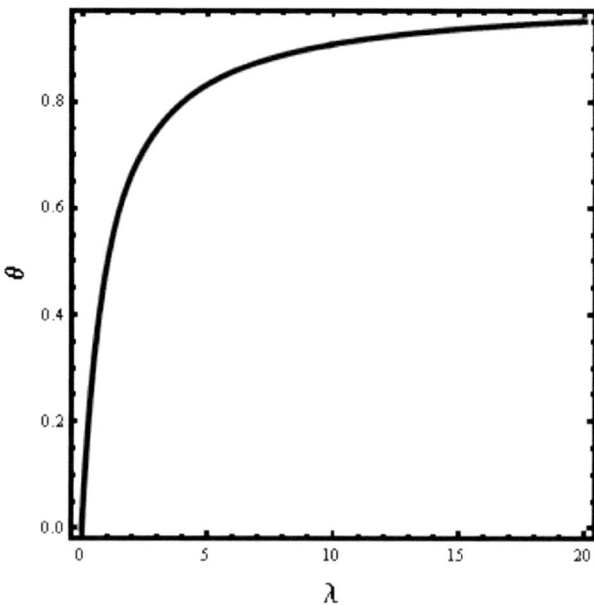

Fig. 7.6 The Langmuir isotherm (Eq. (7.49)).

Note that when there are no nearest-neighbor interactions, i.e. for $q = 1$, we get

$$\theta = \frac{\lambda}{1 + \lambda} \tag{7.50}$$

which is similar to the Langmuir isotherm, Fig. 7.6. with zero interaction energy.

7.3 Lattice Model of a Two-Component Mixture with Nearest Neighbor Interactions

As in the model of the previous section, we have M sites each of which may be occupied by either an A or a B molecule, see Fig. 7.7. We denote by $W_{\alpha\beta}$ the interaction energy between two molecules α and β on adjacent sites. The energy levels of

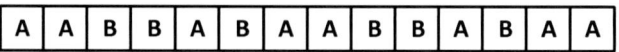

Fig. 7.7 A 1D lattice model of mixture A and B.

the system are determined by

$$E(\mathbf{s}) = N_{AA} W_{AA} + N_{BB} W_{BB} + N_{AB} W_{AB} \qquad (7.51)$$

where $N_{\alpha\beta}$ is the number of nearest-neighbor pair of the type $\alpha\beta$. If N_A and N_B are the number of A and B molecules, and all sites are presumed occupied by either A or B, we must have

$$N_A + N_B = M \qquad (7.52)$$

We also have the two relationships

$$2N_A = 2N_{AA} + N_{AB} \qquad (7.53)$$
$$2N_B = 2N_{BB} + N_{AB} \qquad (7.54)$$

Each AA bond contributes two As; each AB bond contributes one A. If we sum over all N_{AA} bonds and all N_{AB} bonds, we count each A twice; hence, relation (7.52). A similar argument applies for (7.53)

The canonical PF is

$$
\begin{aligned}
Q(T&, M, N_A, N_B) \\
&= \sum_{\substack{\text{all states with} \\ N_A + N_B = M}} \exp[-\beta(N_{AA} W_{AA} + N_{BB} W_{BB} + N_{AB} W_{AB})] \\
&= \exp(-\beta N_A W_{AA}) \exp(-\beta N_B W_{BB}) \sum_{N_{AB}} \Omega(N_{AB}) \\
&\quad \times \exp\left(\frac{-\beta W N_{AB}}{2}\right) \qquad (7.55)
\end{aligned}
$$

where the parameter W is defined by

$$W = 2W_{AB} - W_{AA} - W_{BB} \qquad (7.56)$$

In (7.56) we used relationship (7.52) and (7.53) to rewrite the sum over all states as a sum over all possible values of the parameter N_{AB}. The degeneracy corresponding to the energy level (7.50) is denoted $\mathbf{\Omega}(N_{AB})$. As in the previous model, this degeneracy can be calculated for the one-dimensional model. A more elegant treatment is to use the matrix method again. We open the system with respect to both A and B; the corresponding grand PF is

$$\Xi(T, M, \lambda_A, \lambda_B)$$

$$= \sum_{N_B=0}^{M} \sum_{N_A=0}^{M} Q(T, M, N_A, N_B) \lambda_A^{N_A} \lambda_B^{N_B} \qquad (7.57)$$

The summation in (7.56) is over all N_A and N_B with the condition $N_A + N_B = M$. λ_A and λ_B are the absolute activities of A and B, respectively.

Defining the elements of the matrix P

$$P_{AA} = \langle A|P|A\rangle = \lambda_A \exp\left(-\beta W_{AA}\right) = \lambda_A q_{AA}$$

$$P_{AB} = \langle A|P|B\rangle = \langle B|P|A\rangle = \lambda_A^{1/2}\lambda_B^{1/2} \exp\left(-\beta W_{AB}\right)$$

$$= \lambda_A^{1/2}\lambda_B^{1/2} q_{AB}$$

$$P_{BB} = \langle B|P|B\rangle = \lambda_B \exp\left(-\beta W_{BB}\right) = \lambda_B q_{BB} \qquad (7.58)$$

where all the quantities $q_{\alpha\beta}$ are defined in Eq. (7.58).

We now rewrite the PF (7.56) after adding the factor $\exp\left(-\beta W_{s_N s_1}\right)$ to close the system into a circle, and obtain

$$
\begin{aligned}
\Xi(T, M, &\lambda_A, \lambda_B) \\
&= \sum_{s} \langle s_1|P|s_2\rangle \langle s_2|P|s_3\rangle \cdots \langle s_{M-1}|P|s_M\rangle \langle s_M|P|s_1\rangle \\
&= Tr P^M = \gamma_+^M \qquad\qquad (7.59)
\end{aligned}
$$

where γ_+ is the larger of the two eigenvalue of the matrix P. Thus

$$
\gamma_+ = \frac{1}{2}\{\lambda_A q_{AA} + \lambda_B q_{BB}
$$
$$
+ [(\lambda_A q_{AA} - \lambda_B q_{BB})^2 + 4\lambda_A\lambda_B q_{AB}^2]^{1/2}\} \qquad (7.60)
$$

From Eqs. (7.58) and (7.59) we can calculate all relevant thermodynamic quantities of the system.

A simple case occurs when $W = 0$, i.e. when W_{AB} is the arithmetic average of W_{AA} and W_{BB}, or equivalently

$$
q_{AB}^2 = q_{AA} q_{BB} \qquad\qquad (7.61)
$$

In this case γ_+ reduces to

$$
\gamma_+ = \lambda_A q_{AA} + \lambda_B q_{BB} \qquad\qquad (7.62)
$$

This case is referred to as the symmetrical ideal solution. The average number of As (or Bs) in the system is obtained from

$$
\bar{N}_A = \lambda_A M \frac{\partial \ln \gamma_+}{\partial \lambda_A} \qquad\qquad (7.63)
$$

The expression for $\bar{N}_A$ is quite cumbersome in the general case. However, it is very simple in the ideal case (7.60), for which γ_+ is given by (7.61), hence

$$\bar{N}_A = M \frac{\lambda_A q_{AA}}{\lambda_A q_{AA} + \lambda_B q_{BB}} \qquad (7.64)$$

Denote by x_A the mole fraction of A in the system, we have

$$x_A = \frac{\lambda_A q_{AA}}{\lambda_A q_{AA} + \lambda_B q_{BB}} \qquad (7.65)$$

and a similar expression for

$$x_B = (1 - x_A) = \frac{\lambda_B q_{BB}}{\lambda_A q_{AA} + \lambda_B q_{BB}} \qquad (7.66)$$

The ratio of $\bar{N}_A$ and $\bar{N}_B$ is thus

$$\eta = \frac{\bar{N}_A}{\bar{N}_B} = \frac{\lambda_A q_{AA}}{\lambda_B q_{BB}}$$
$$= \exp[\beta(\mu_A - \mu_B)] \exp[-\beta(W_{AA} - W_{BB})] \qquad (7.67)$$

This ratio is controlled by the difference in the chemical potentials $\mu_A - \mu_B$, and by the difference $W_{AA} - W_{BB}$. For any fixed values of the molecular parameters $W_{\alpha\beta}$, the ratio η will tend to infinity when $\lambda_A \gg \lambda_B$, and tend to zero when $\lambda_A \ll \lambda_B$.

In the more general (non-ideal) case one can perform the derivatives (7.62) and, after some algebra, obtain the final results

$$x_A = \frac{\bar{N}_A}{N} = \frac{\lambda_B q_{BB} - \gamma_+}{\lambda_A q_{AA} + \lambda_B q_{BB} - 2\gamma_+} \qquad (7.68)$$

$$x_B = \frac{\bar{N}_B}{N} = \frac{\lambda_A q_{AA} - \gamma_+}{\lambda_A q_{AA} + \lambda_B q_{BB} - 2\gamma_+} \qquad (7.69)$$

which provides the dependence of the concentrations of A and B on the absolute activities, as well as on the molecular parameter of the model.

7.4 Two States of a Polymer Having Different Lengths

As the final example of a system consisting of two-state particles, we consider a linear string of polymers. Each polymer can be in one of two states, long (A) or short (B) with corresponding lengths l_A and l_B, respectively, see Fig. 7.8. We denote by Q_A and Q_B the internal PF of the polymer in the A and B state, and by W_{AA}, W_{BB} and W_{AB} the corresponding interaction between nearest pairs. The total length of the system is

$$L = N_A l_A + N_B l_B \qquad (7.70)$$

Note also that we assume that $W_{AB} = W_{BA}$. If the two ends of the polymer are not identical, or similar, then we must distinguish between W_{AB} and W_{BA}.

Since $N = N_A + N_B$ is fixed, then fixing L also fixes N_A, by (7.70). Therefore, the canonical PF for this system is

$$Q(T, L, N)$$
$$= Q_A^{N_A} Q_B^{N_B} \sum$$
$$\times \exp[-\beta(N_{AA}W_{AA} + N_{BB}W_{BB} + N_{AB}W_{AB})] \qquad (7.71)$$

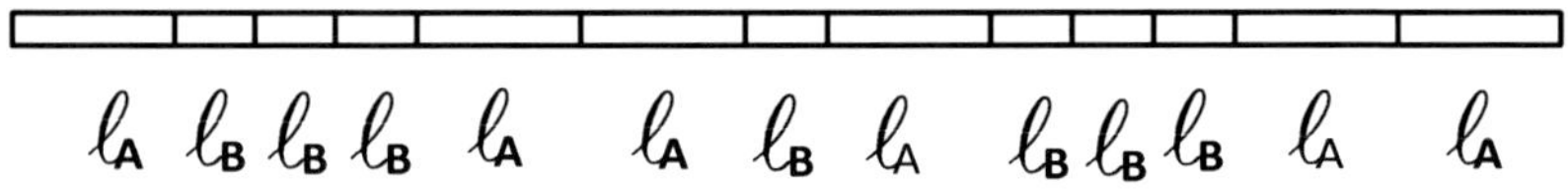

Fig. 7.8 A 1D model of groups of different lengths.

where the sum is over all possible configurations with fixed values of N_A and N_B. In order to remove the condition of fixed N_A, or equivalently, of fixed total length L, we transform the variable L into the corresponding one-dimensional pressure, or tension τ.

$$\Delta(T, \tau, N) = \sum_L Q(T, L, N) \exp(\beta\tau L) \qquad (7.72)$$

where the sum is over all possible lengths of the system. If we choose, say $l_A > l_B$, then the minimum of the value of L is Nl_B, and the maximum value is Nl_A.

The tension τ is defined as positive when the system is stretched. In the three-dimensional system, the pressure is defined as positive when the system is compressed. Therefore, we have the factor $\exp(-\beta PV)$ in the T, P, N ensemble, whereas here we have $\exp[\beta\tau L]$.

We now define the four elements of the 2×2 matrix P:

$$P_{AA} = \langle A|P|A \rangle = Q_A\delta_A \exp(-\beta W_{AA}) = Q_A\delta_A q_{AA}$$
$$P_{AB} = P_{BA}\langle A|P|B \rangle = (Q_A\delta_A)^{1/2}(Q_B\delta_B)^{1/2} \exp(-\beta W_{AB})$$
$$= (Q_A Q_B \delta_A \delta_B)^{1/2} q_{AB}$$
$$P_{BB} = \langle B|P|B \rangle = Q_B\delta_B \exp[-\beta W_{BB}] = Q_B\delta_B q_{BB}$$

$$(7.73)$$

where

$$\delta_A = \exp(\beta\tau l_A), \quad \delta_B = \exp(\beta\tau l_B).$$

With these definitions we can perform the sum in (7.22) simply by writing one specific configuration and summing over all possible configurations. The sum over all L values is the same as the sum over all possible vectors s of the system.

The result after closing the system to a cycle is:

$$\Delta(T, \tau, N) = \sum_s \langle s_1|P|s_1\rangle \cdots \langle s_N|P|s_1\rangle = Tr P^N \quad (7.74)$$

The larger of the two eigenvalues of P can be obtained as before, the result being

$$\gamma_+ = \frac{1}{2}\{(P_{AA} + P_{BB}) + [(P_{AA} - P_{BB})^2 + 4P_{AB}^2]^{1/2}\}$$

$$= \frac{1}{2}\{(Q_A\delta_A q_{AA} + Q_B\delta_B q_{BB}) + [(Q_A\delta_A q_{AA} - Q_B\delta_B q_{BB})^2$$

$$+ 4Q_A Q_B \delta_A \delta_B q_{AB}^2]^{1/2}\} \quad (7.75)$$

The analogy with the previous models should be noted. Here, the tension τ plays the role of the activity or the magnetic field in the previous models.

The average length of the system is obtained from

$$\bar{L} = k_B T \frac{\partial \ln \Xi}{\partial \tau} = N k_B T \frac{\partial \ln \gamma_+}{\partial \tau} \quad (7.76)$$

The average length per particle can be put in the form

$$\bar{l} = \frac{\bar{L}}{N} = x_A l_A + x_B l_B \quad (7.77)$$

where

$$x_A = \frac{Q_B\delta_B q_{BB} - \gamma_+}{Q_A\delta_A q_{AA} + Q_B\delta_B q_{BB} - 2\gamma_+}, \quad x_B = 1 - x_A \quad (7.78)$$

The equilibrium constant for the conversion $A \rightleftarrows B$ is

$$\eta = \frac{x_A}{x_B} = \frac{Q_B\delta_B q_{BB} - \gamma_+}{Q_A\delta_A q_{AA} - \gamma_+} \quad (7.79)$$

A simple case occurs when $2W_{AB} = W_{AA} + W_{BB}$ or $q_{AB}^2 = q_{AA}q_{BB}$. (This corresponds to the symmetrical ideal solution of the previous section). In this case,

$$\gamma_+ = Q_A \delta_A q_{AA} + Q_B \delta_B q_{BB} \tag{7.80}$$

and the equilibrium constant η reduces to

$$\eta = \frac{x_A}{x_B} = \frac{Q_A \delta_A q_{AA}}{Q_B \delta_B q_{BB}} = \frac{Q_A q_{AA}}{Q_B q_{BB}} \exp[\beta\tau(l_A - l_B)] \tag{7.81}$$

If we choose $l_A - l_B > 0$, then increasing the tension favors the longer form, A. Figure 7.9 shows how the ratio η changes with T. In the more general case Eq. (7.78) provides the dependence of the equilibrium constant on the tension, as well as on the molecular parameters of the model.

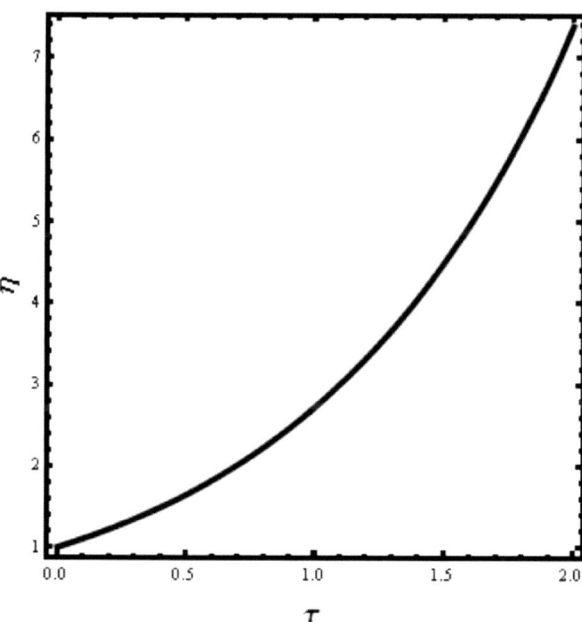

Fig. 7.9 The change of the length of the system in Fig. 7.8 with the tension τ.

8

Theory of a One-Dimensional Pure and Simple Fluid

This chapter is devoted to the study of some basic properties of 1D pure simple fluid. The reader should be aware of the fact that while there are exact theories of ideal gases, as well as deviation from ideal gases, and exact theories of simple solids, there is no such theory for any 3D liquid, not even for the simplest liquid composed of hard particles.

As we have emphasized in the introduction to the previous chapter, the study of 1D system is important not only for its applications, but also for the very method of statistical mechanics to obtain exact solutions to non-trivial models. As in the case of lattice models discussed in Chapter 7, there is an exact solution to the 1D fluid. The most important application of these models is in the study of the properties of liquid water, which is discussed in Chapter 10. In this chapter we present the method of solving for the partition function of a one-component fluid in a 1D system.

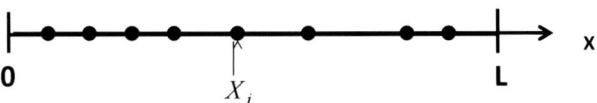

Fig. 8.1   A 1D fluid of N particles in a "box" of length L.

8.1 The Partition Function of the 1D Fluid

Consider a system of N particles in a one-dimensional "box" of length L. The particles are free to move within the limits $(0, L)$ of the system. The location of the center of the ith particle is denoted by X_i. See Fig. 8.1.

The total interaction energy of the system is assumed to be the sum of the nearest-neighbor pair interactions, i.e.,

$$U_N(X_1, \ldots, X_N) = \sum_{i=1}^{N-1} U(X_i, X_{i+1}) \qquad (8.1)$$

where $U(X_i, X_{i+1})$ is the interaction potential between two consecutive particles i and $i+1$. We shall always assume that this potential function is a function of the distance between particles i and $i + 1$; $X_{i,i+1} = X_{i+1} - X_i$. The function $U(X)$ is also assumed to be zero at sufficiently large distance, and infinitely repulsive at very small distances. Three possible forms of the function $U(X)$ are shown in Fig. 8.2.

The canonical PF of the system is written as

$$Q(T, L, N)$$
$$= \frac{1}{N! \Lambda^N} \int_0^L \cdots \int_0^L dX_1 \cdots dX_N \exp\left(-\beta U_N\right)$$
$$(8.2)$$

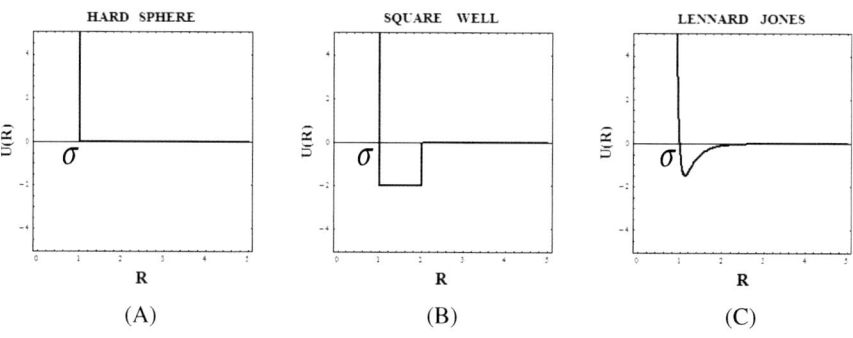

Fig. 8.2 Three possible pair potentials.

where Λ is the momentum *PF*, or the de-Broglie wavelength of the particles in one dimension, defined by

$$\Lambda = \frac{h}{\sqrt{2\pi m k_B T}} \qquad (8.3)$$

This quantity is obtained from the integrations over all possible momenta of the particles in the system. h is the Planck constant, m is the mass of the particles, and k_B the Boltzmann constant. U_N is a shorthand notation for the total potential energy defined in Eq. (8.1).

In (8.2), the range of integration for each particle is from 0 to L. This means that any configuration $X_1, \ldots, X_N$ is possible. However, in a 1D system, the particles can be *ordered*, in the sense that $X_1 < X_2 < X_3 \cdots < X_N$. This allows us to rewrite (8.2) in a modified form as follows: For any function $f(X_1, \ldots, X_N)$, symmetric with respect to interchanging the positions of any of the particles X_i and X_j

we can write the identity

$$\int_0^L \cdots \int_0^L dX_1 \cdots dX_N f(X_1, \ldots, X_N)$$

$$= N! \int_0^L dX_N \int_0^{X_N} dX_{N-1} \cdots \int_0^{X_2} dX_1 f(X_1, \ldots, X_N)$$

$$(8.4)$$

This can be easily verified for small N. For instance, for $N = 2$ we have

$$\int_0^L \int_0^L dX_1 dX_2 f(X_1, X_2) = \iint_{X_1 > X_2} + \iint_{X_1 < X_2} \qquad (8.5)$$

The total region of integration is the square area of Fig. 8.3. This region is split into two; above and below the diagonal line. The integral on the lhs of (8.5) is rewritten as the sum of two integrals; each over one of the triangled area of Fig. 8.3. $f(X_1, X_2)$ is presumed to be symmetric with respect to the two variables X_1, X_2. Therefore, for each point in one triangle, there is corresponding point in the second triangle for which

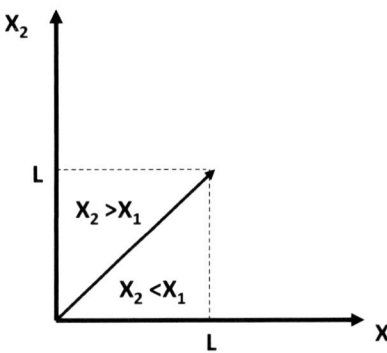

Fig. 8.3 Two regions of integration, see Eq. (8.5).

the function f has the same value. Therefore, instead of integrating over the entire squared area, it is sufficient to take two times the integral over one of the triangles, i.e.,

$$\int_0^L \int_0^L dX_1 dX_2 f(X_1, X_2) = 2 \int_{X_1 < X_2} dX_1 dX_2 f(X_1, X_2)$$

$$= 2 \int_0^L dX_2 \int_0^{X_2} dX_1 f(X_1, X_2) \tag{8.6}$$

For the case $N = 3$ we have six volumes within the cube of edge L. The six regions are defined $0 < X_1 < X_2 < X_3 < L$, $0 < X_1 < X_3 < X_2 < L$, etc. See Fig. 8.4.

The reader should be aware of the fact that the "ordering" of the particles is possible in 1D system. It cannot be done either in 2D or 3D systems. This is the main reason for the solvability of the PF of the 1D fluid.

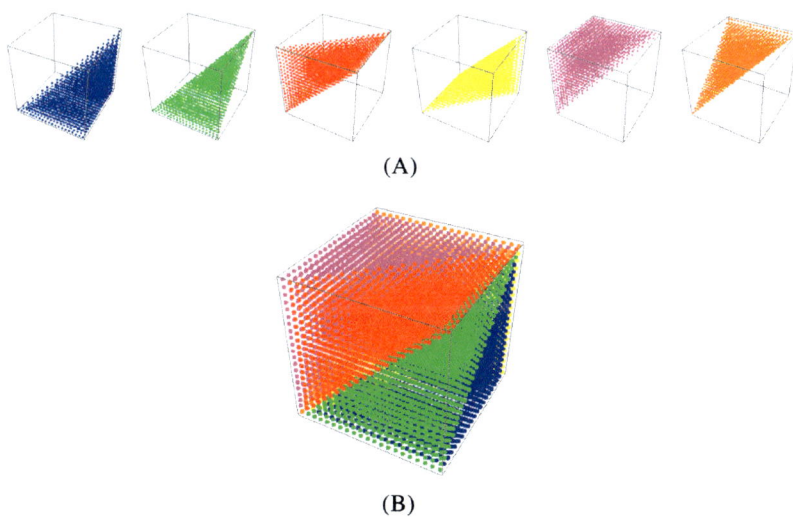

Fig. 8.4 The six regions of integrations for three particles in 1D system. (A) separate regions (B) all the regions.

Because of the particles are indistinguishable, the function $U(X_1, \cdots, X_N)$ is symmetric with respect to interchanging the positions of any two particles. Therefore, we can apply the identity (8.4) to the integrand $\exp(-\beta U_N)$ in (8.2) and rewrite the canonical *PF* of the system as:

$$Q(T, L, N)$$
$$= \frac{1}{\Lambda^N} \int_0^L dX_N \int_0^{X_N} dX_{N-1} \cdots \int_0^{X_2} dX_1 \exp(-\beta U_N)$$
$$(8.7)$$

Note carefully that the factor $N!$ in (8.2) is added to the PF to correct for over-counting of configurations. By fixing the *order* of the particles, we actually labeled the particles as: First, second, third . . . and Nth. This labeling makes the particles distinguishable; hence, the factor $N!$ is removed in (8.7).

Because of the ordering of the particles and because of the assumption of pair-wise additivity of the total interactions in (8.1), we can rewrite the PF in simpler form by taking the Laplace transform and applying the convolution theorem. In our case, this is equivalent to transforming the canonical PF in (8.7) to the PF in the T, P, N ensemble. See Chapter 1. Here P is the "pressure" in the one-dimensional system.

Taking the Laplace transform of $Q(T, L, N)$ with respect to the variable L, we get

$$\Delta(T, P, N) = C \int_0^\infty Q(T, L, N) \exp(-\beta PL) dL \quad (8.8)$$

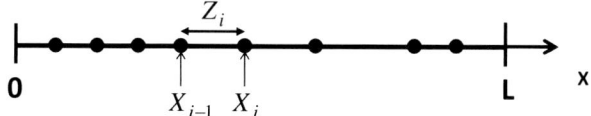

Fig. 8.5 Definition of the distances $Z_i = X_i - X_{i-1}$.

To simplify the final form of the *PF*, we add a zeroth particle at the origin X_0 and an $(N+1)$th particle at $X_{N+1} = L$. C is a constant having the dimensions of L^{-1}, so that the entire rhs of (8.8) became dimensionless. Since this constant does not affect the thermodynamic properties of the system, we shall simply take $C = 1$.

We now transform to relative coordinates, i.e. to the distances between two consecutive particles. We define the distances. See Fig. 8.5.

$$Z_1 = X_1 - X_0 = X_1 - 0$$
$$Z_i = X_i - X_{i-1}$$
$$Z_{N+1} = L - X_N \tag{8.9}$$

Next, we rewrite the total interaction energy as

$$U(X_0, \ldots, X_{N+1}) = \sum_{i=1}^{N+1} U(Z_i) \tag{8.10}$$

The Jacobian of the transformation of variables (8.9) is unity, i.e.

$$\frac{\partial(X_1, \ldots, X_{N+1})}{\partial(Z_1, \ldots, Z_{N+1})} = 1 \tag{8.11}$$

Therefore, we can rewrite the *PF* in (8.8) as

$$\Delta(T, P, N) = \frac{1}{\Lambda^N} \int_0^\infty dZ_{N+1} \int_0^\infty dZ_N \cdots \int_0^\infty dZ_1$$

$$\times \exp\left[-\beta P \sum_{i=1}^{N+1} Z_i - \beta \sum_{i=1}^{N+1} U(Z_i)\right]$$

(8.12)

In (8.12) we wrote the length of the system L as the sum of the distances Z_i, i.e.,

$$L = \sum_{i=1}^{N+1} Z_i \qquad (8.13)$$

The integrand in (8.12) is a product of $N + 1$ factors, each depends only on one of the variables Z_i. Therefore, the $(N + 1)$-dimensional integral can be rewritten as a product of $N + 1$ one-dimensional integrals, i.e.

$$\Delta(T, P, N) = \frac{1}{\Lambda^N} \left\{\int_0^\infty dR \, \exp[-\beta PR - \beta U(R)]\right\}^{N+1}$$

(8.14)

We now define the function

$$\Psi(P, T) = \int_0^\infty dR \, \exp[-\beta PR - \beta U(R)] \qquad (8.15)$$

For large number of particles $N \gg 1$, we can rewrite the *PF* in (8.14) as

$$\Delta(T, P, N) = \frac{\Psi(P, T)^N}{\Lambda^N} \qquad (8.16)$$

Note that in (8.16) we ignored one factor $\Psi(P, T)$. This is equivalent to *closing* the one-dimensional system, so that the particle at $X_{N+1} \equiv L$ is identified with the particle at $X_0 = 0$. Thus, for any given pair potential $U(R)$, we can calculate the quantity $\Psi(P, T)$ in (8.15), and hence the *PF* of the system in (8.16). From the *PF* we can derive all the thermodynamic quantities for the system.

8.2 Thermodynamics of the 1D Fluid

The Gibbs energy is obtained from the relationship

$$G(T, P, N) = -k_B T \ln \Delta$$
$$= N k_B T \ln \Lambda - N k_B T \ln \Psi(P, T) \quad (8.17)$$

The average length of the system is obtained from the thermodynamic relationship

$$\bar{L} = \frac{\partial G}{\partial P} = N \frac{\int_0^\infty dR \exp[-\beta PR - \beta U(R)]R}{\int_0^\infty dR \exp[-\beta PR - \beta U(R)]}$$
$$= N \langle R \rangle = N \int_0^\infty Pr(R) R dR \quad (8.18)$$

In (8.18), we wrote the average length of the system as N times the average length per particle $\langle R \rangle$. Note that this average is taken with the distribution function

$$Pr(R) = \frac{\exp[-\beta PR - \beta U(R)]}{\int_0^\infty dR \exp[-\beta PR - \beta U(R)]} \quad (8.19)$$

$Pr(R)$ is the probability density of finding a pair of nearest neighbors at a distance R.

From (8.18) and (8.19) we can derive the expression for the isothermal compressibility

$$\kappa_T = \frac{-1}{\bar{L}}\left(\frac{\partial \bar{L}}{\partial P}\right)_T$$

$$= \frac{\beta[\langle R^2\rangle - \langle R\rangle^2]}{\langle R\rangle} = \frac{\beta\langle(R - \langle R\rangle)^2\rangle}{\langle R\rangle} > 0 \qquad (8.20)$$

Thus, as expected, the isothermal compressibility is always positive.

The thermal expansion coefficient is given by

$$\alpha_p = \frac{1}{\bar{L}}\left(\frac{\partial \bar{L}}{\partial P}\right)_T = \frac{\langle RH(R)\rangle - \langle R\rangle\langle H(R)\rangle}{k_B T^2\langle R\rangle} \qquad (8.21)$$

where we have defined by $H(R) = U(R) + PR$. This quantity will be important in the study of liquid water, which is explained in Chapter 10.

The chemical potential may be obtained directly from (8.17), by taking the derivative with respect to N.

$$\mu = k_B T \ln \Lambda - k_B T \ln \Psi(P, T) \qquad (8.22)$$

The entropy of the system is obtained from the temperature derivative of the Gibbs energy

$$S = -\left(\frac{\partial G}{\partial T}\right)_{P,N} = -Nk_B \ln \Lambda - Nk_B T\frac{\partial \ln \Lambda}{\partial T}$$

$$+ Nk_B \ln \psi + Nk_B T\frac{\partial \ln \psi}{\partial T} \qquad (8.23)$$

Using (8.16) and (8.19) we can rewrite the entropy as

$$S = -Nk_B \ln \Lambda - Nk_B T \frac{\partial \ln \Lambda}{\partial T}$$

$$- Nk_B \int_0^\infty dR Pr(R) \ln Pr(R) \qquad (8.24)$$

Note the form of the third term on the rhs of (8.24). This is the Shannon measure of information associated with the distribution $Pr(R)$.

8.3 Ideal Gas in a 1D System

A special case of a 1D fluid is when there are no interactions between the particles, i.e. $U(R) = 0$. In this case we have an ideal gas, for which

$$\Psi(P, T) = (\beta P)^{-1} \qquad (8.25)$$

Hence, the chemical potential is

$$\mu = k_B T \ln (\Lambda \beta P) \qquad (8.26)$$

The equation of state of an ideal gas is

$$\bar{l} = \left(\frac{\partial \mu}{\partial P}\right)_T = \frac{k_B T}{P} \qquad (8.27)$$

or equivalently

$$\rho = \beta P \qquad (8.28)$$

where $\rho = N/V$ is the density of the particles. Note that this equation is the same as the equation of state for the 3D ideal gas.

8.4 A System of Hard Rods

Consider a system of hard rods, each of length σ. The pair potential for such particles is (See Fig. 8.2a):

$$U(R) = \begin{cases} \infty & R \leq \sigma \\ 0 & R > \sigma \end{cases} \qquad (8.29)$$

where σ is the length of the particles. For this case we have

$$\Psi(P, T) = \int_{\sigma}^{\infty} dR \; \exp(-\beta PR) = \frac{\exp(-\beta P\sigma)}{\beta P} \qquad (8.30)$$

The chemical potential is

$$\mu(P, T) = k_B T \ln (\Lambda \beta P) + \sigma P \qquad (8.31)$$

and the equation of state is

$$P\bar{L} = Nk_B T + N\sigma P \qquad (8.32)$$

Note that from (8.32) it follows that the minimum value of the average volume is obtained in the limit $P \to \infty$ (Fig. 8.6).

$$\bar{L}_{min} = \lim_{p \to \infty} \bar{L} = N\sigma \qquad (8.33)$$

This corresponds to the maximum density of σ^{-1}, which is the closest packing density.

Fig. 8.6 Maximum density in the 1D system.

Denoting by $\rho = N/\bar{L}$ the average density in the T, P, N ensemble, we can write the equation of state (8.32) as

$$\frac{P}{k_B T} = \frac{\rho}{1 - \rho\sigma} = \rho[1 + \rho\sigma + (\rho\sigma)^2 + \cdots] \qquad (8.34)$$

This power series converges for $\rho\sigma < 1$. Note that when $\rho\sigma \geq 1$, the average density is larger than σ^{-1}. This is impossible because of (8.33), i.e. the density cannot exceed the maximum density of σ^{-1}.

The terms in the square brackets in Eq. (8.34) give the successive corrections to the ideal gas pressure due to interactions between pairs, triplets, etc. of particles. This expansion is the virial expansion of the pressure for the one dimensional system.

Figure 8.7 shows the equation of a state $P(\rho)$ for different values of T and $k_B = 1, \sigma = 1$.

Note that the slope of $P/k_B T$ vs. ρ curve is always positive, i.e.

$$\frac{\partial (P/k_B T)}{\partial \rho} = \frac{1}{(1 - \rho\sigma)^2} > 0 \qquad (8.35)$$

This means that there could be no phase transition in the system. The isothermal compressibility of the system is obtained from (8.32) and (8.34)

$$\kappa_T = \frac{-1}{\bar{l}}\left(\frac{\partial \bar{l}}{\partial P}\right)_T = \frac{(1 - \rho\sigma)^2}{\rho k_B T} > 0 \qquad (8.36)$$

This is always a positive quantity (Fig. 8.8).

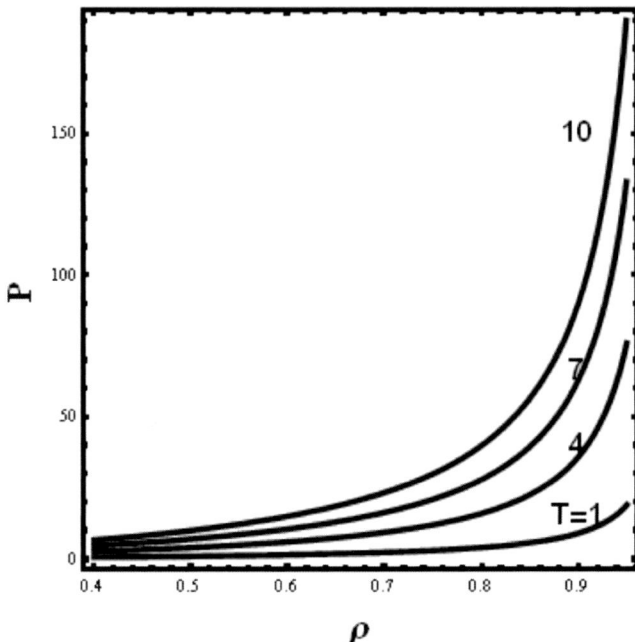

Fig. 8.7 The equation of state $P(\rho)$, Eq. (8.34) at different temperatures T.

Another quantity that we shall find useful in mixtures is the pseudo-chemical potential. This is defined, in general as

$$\mu^* = \mu - k_B T \ln \rho \Lambda \qquad (8.37)$$

For the case of hard rods, we can first express the chemical potential in (8.31) in the form

$$\mu = k_B T \ln \rho \Lambda - k_B T \ln (1 - \rho \sigma) + \frac{k_B T \rho \sigma}{1 - \rho \sigma} \qquad (8.38)$$

The pseudo-chemical potential has a simple interpretation. This is the change of the Gibbs energy for introducing one particle at some fixed position in the system. From (8.38) we extract the pseudo-chemical potential for a

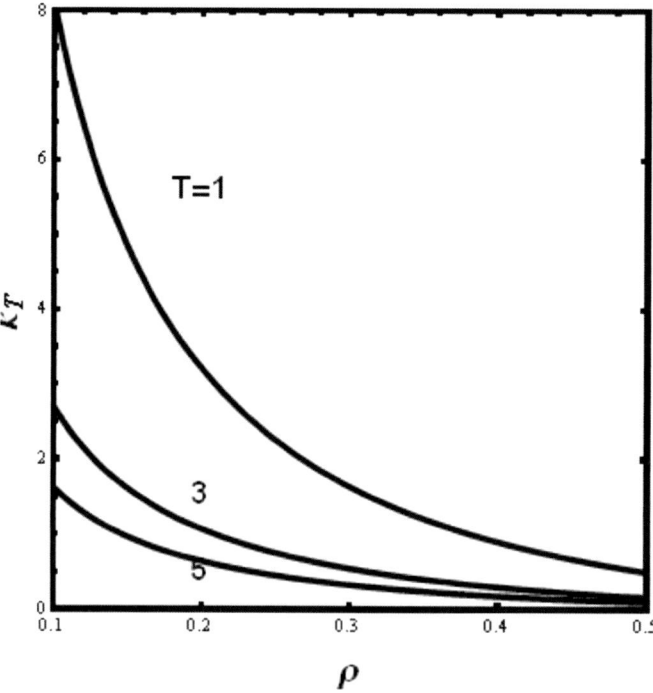

Fig. 8.8 The isothermal compressibility κ_T as a function of the density ρ, at different temperatures.

one-component of hard rods system

$$\mu^* = -k_B T \ln (1 - \rho\sigma) + \frac{k_B T \rho\sigma}{1 - \rho\sigma}$$

$$= k_B T \ln (1 + \beta P\sigma) + P\sigma \tag{8.39}$$

The second form on the rhs of (8.39) gives μ^* as a function of T and P and is useful for calculating the thermodynamics of solvation.

The Gibbs energy of solvation is defined as

$$\Delta G^* = \mu^{*l} - \mu^{*ig}$$

$$= k_B T \ln (1 + \beta P\sigma) + P\sigma \tag{8.40}$$

Note that since hard rods do not have internal degrees of freedom the pseudo-chemical potential of an ideal gas (*ig*) is zero.

Other solvation thermodynamic quantities in this system are:

$$\Delta S^* = -\left(\frac{\partial \Delta G^*}{\partial T}\right)_P$$

$$= -k_B \ln(1 + \beta P \sigma) + \frac{1}{T} \frac{P\sigma}{(1 + \beta P \sigma)} \quad (8.41)$$

$$\Delta H^* = \Delta G^* + T \Delta S^* = P\sigma + \frac{P\sigma}{1 + \beta P \sigma} \quad (8.42)$$

$$\Delta V^* = \left(\frac{\partial \Delta G^*}{\partial P}\right)_T = \sigma + \frac{\sigma}{1 + \beta P \sigma} \quad (8.43)$$

$$\Delta E^* = \Delta H^* - P \Delta V^* = 0 \quad (8.44)$$

Note that for $P \to 0$

$$\Delta V^* = 2\sigma \quad (8.45)$$

This may be interpreted as the excluded volume of each particle. See Fig. 8.9.

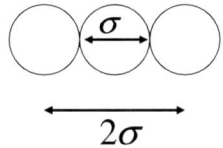

Fig. 8.9   The excluded volume (2σ) about a particle with diameter σ.

whereas for $P \rightarrow \infty$

$$\Delta V^* = \sigma \qquad (8.46)$$

This is the change in volume of the system when $P \rightarrow \infty$, which is consistent with the limit of the volume per particle (8.33).

Note that if we take the limit $P \rightarrow 0$ in (8.39) we get $\mu^* = 0$, the same is true for (8.41). In this case we shall not get the correct values of the solvation entropy, volume, etc.

Figure 8.10 shows the solvation Gibbs energy of a hard rod in a hard rod system, $(\sigma = 1, k_B = 1)$ as a function of pressure for different temperatures.

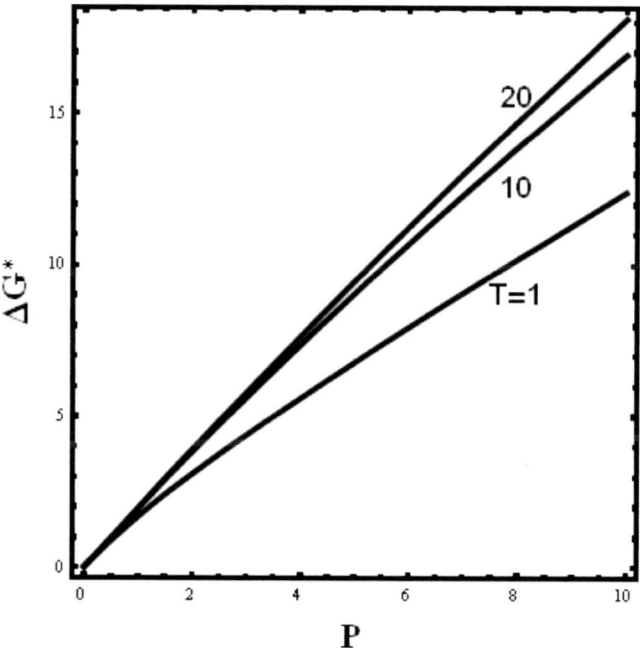

Fig. 8.10 The solvation Gibbs energy of hard rods system as a function of the pressure at different temperatures.

8.5 System of Particles Interacting via Square Well Potentials

The next system we shall study is a system of particles interacting via a pair potential of the form

$$U(R) = \begin{cases} \infty & 0 \leq R \leq \sigma \\ -\varepsilon & \sigma < R \leq 2\sigma \\ 0 & R > 2\sigma \end{cases} \quad (8.47)$$

We will refer to these particles as square well (SW) particles, see Fig 8.2b.

We first show in Fig. 8.11 the probability density function $Pr(R)$ for different values of T, P and ε. Note that $\varepsilon = 0$ corresponds to the hard rods system.

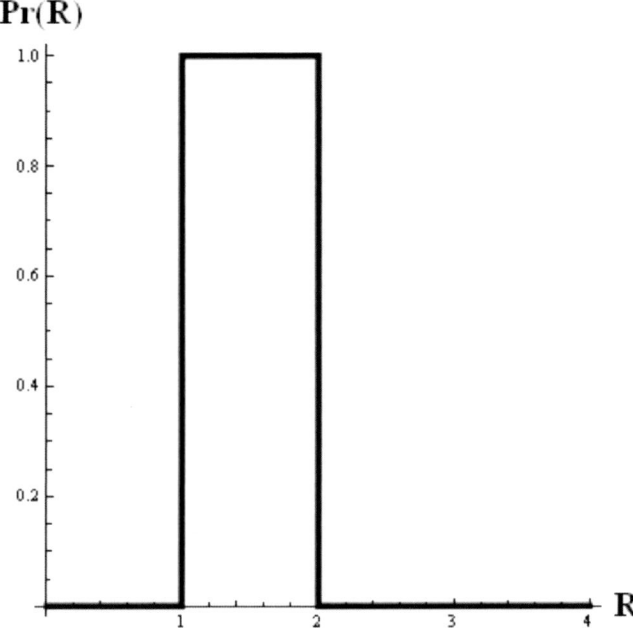

Fig. 8.11 The distribution Pr(R), Eq. (8.19) for square well potential.

The chemical potential for a one-component system may be written as:

$$\mu = k_B T \ln \rho \Lambda - k_B T \ln \rho \Psi(P, T) \tag{8.48}$$

The general expression for the pseudo-chemical potential for a pure 1D fluid is

$$\mu^*(P, T) = -k_B T \ln \rho \Psi(P, T) \tag{8.49}$$

where $\rho = \rho(P, T)$ is the average density in the T, P, N system.

To express μ^* as a function of T, P we first note that

$$\left(\frac{\partial \psi}{\partial P}\right)_T = \int_0^\infty dR \, \exp[-\beta PR - \beta U(R)] \, (-\beta R) \tag{8.50}$$

Hence, the density $\rho(P, T)$ can be obtained from (8.18) and (8.50)

$$\rho(P, T) = \frac{-\beta \psi(P, T)}{\psi'(P, T)} \tag{8.51}$$

where $\psi'(P, T) = (\partial \psi / \partial P)_T$. Substituting (8.51) into (8.49) we get

$$\mu^*(P, T) = -k_B T \ln \left[\frac{-\beta \psi^2(P, T)}{\psi'(P, T)} \right] \tag{8.52}$$

from which can be obtained all the thermodynamic quantities of solvation in pure system. We shall generalize this equation for mixtures in the next chapter.

9

Theory of 1D Dimensional Mixtures

This chapter is an extension of the previous one; from a one-component fluid to a multi-component fluid. Traditionally, there has been a distinction between *mixtures* and *solutions*. In the latter, we distinguish between one (or more) components, referred to as a solute dissolved in a second component, referred to as the solvent. In the former, the two (or more) components are treated more symmetrically. Today we need not make this distinction. We can use the term "mixtures" for any multi-component system. We will take the limit of low concentrations of one (or more) components, whenever necessary.

In this chapter, we will also encounter a new partition function, which we will refer to as the *generalized* PF. Its application is somewhat non-conventional. However, it is a very useful PF. We will encounter an important application of this PF in the study of aqueous solutions in the next chapter.

9.1 (***) The Partition Function of a 1D Mixture of a *c*-Components System

We start with the canonical *PF* of a system of a *c*-component system in 1D:

$$Q(T, L, N) = \frac{1}{N! \prod_i \Lambda_i^{N_i}}$$

$$\times \int_0^L \cdots \int_0^L \prod_i dX^{N_i} \exp[-\beta U(N)] \quad (9.1)$$

where we used the shorthand notations: $N = N_1, \ldots, N_c$, $N! = \prod_i N_i!$ and $\prod_i dX^{N_i} = dX_1, \ldots, dX_N$, where $N = \sum_{i=1}^c N_i$. As in the previous chapter we assume that the total interaction energy among all the particles in the system is the sum of the interactions between nearest neighbors, i.e.

$$U_N(X_1, \ldots, X_N) = \sum_{i=1}^{N-1} U_{\alpha_i, \alpha_{i+1}}(X_i, X_{i+1}) \quad (9.2)$$

Here, α_i, α_{i+1} is the pairwise interaction energy between particles i and $i+1$ at X_i and X_{i+1}, of species α_i and α_{i+1}, respectively.

In the one-component system we have *ordered* the particles along the "volume" of the system, in such a way that $0 < X_1 < X_2 < X_3 < \cdots < X_N < L$. In our case, since there are *c* different species, we have to distinguish between a *specific ordering* of the *species* (S.O.O.S.) and a specific ordering of the particles (S.O.O.P.). This is an important step we must take before we can apply the convolution theorem to the PF. We first show a simple example of such an ordering.

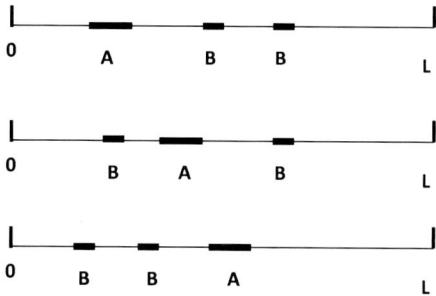

Fig. 9.1 Patterns of interactions between different species for a system of one A and two Bs.

In Fig. 9.1 we have three particles, one A and two Bs is (a and b are the coordinates of A and B, respectively). We start with an arbitrary configuration. We first *order* the species, then we *order* the particles of each species. Formally we write:

$$\int_0^L \int_0^L \int_0^L \exp[-\beta U(a, b_1, b_2)]\, da\, db_1\, db_2$$

$$= \int_0^L \int_{0 \atop ABB}^L \int_0^L \exp[-\beta U_{AB}(a, b_1)$$

$$-\beta U_{BB}(b_1, b_2)]\, da\, db_1\, db_2$$

$$+ \int_0^L \int_{0 \atop BAB}^L \int_0^L \exp[-\beta U_{BA}(b_1, a)$$

$$-\beta U_{AB}(a, b_2)]\, da\, db_1\, db_2$$

$$+ \int_0^L \int_{0 \atop BBA}^L \int_0^L \exp[-\beta U_{BB}(b_1, b_2)$$

$$-\beta U_{BA}(b_2, a)] \, da \, db_1 db_2$$

$$= 2 \int da \int db_1 \int_{ABB} db_2 + 2 \int db_1 \int_{BAB} da \int db_2$$

$$+ 2 \int db_1 \int_{BBA} db_2 \int da \qquad (9.3)$$

Note carefully that in the configurational PF, the first term on the rhs of (9.3), the integration over a, b_1, b_2 is from 0 to L. This means that the specific *order* of the species shown in Fig. 9.1 changes in the course of the integration. Since the integration is over *all* possible configurations, the ordering of the species changes while we perform the integration. In this particular example we have three different orderings of the species: ABB, BAB and BBA. For each of these orderings we have a different integral. We have denoted the specific order of the species under each of the integrals.

Once we fix the *ordering* of the *species* we proceed with the ordering of the particles of each species. This step is the same as we have done in the previous chapter. Recall that by ordering the particles we limit the range of integration for each particle. In this step, unlike in the previous one, the *value* of the integrand does not change upon interchanging particles of the same species. We then multiply each integral by 2 to obtain the last form on the rhs of Eq. (9.3).

The generalization of this procedure for the PF in (9.1) is not easy to visualize. The reader is urged to try some examples like the one in Fig. 9.1 before accepting the general procedure for any number of species and any number of particles.

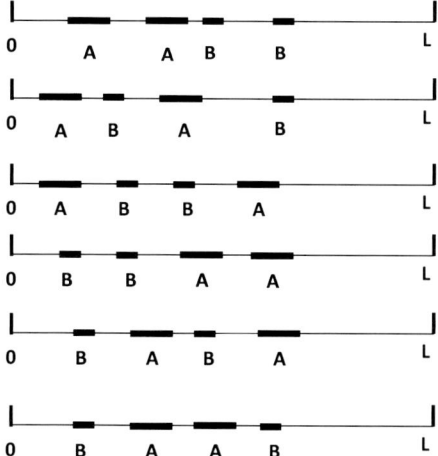

Fig. 9.2 Different ordering of species for four particles: two As and two Bs.

Exercise: Follow the steps as in (9.3) for the case of four particles; AABB. See Fig. 9.2.

The general procedure to convert the configurational integral in (9.1) is

$$\int_0^L \cdots \int_0^L = \sum_{\substack{\text{all orderings} \\ \text{of species} \\ \text{with fixed } N_1,\ldots,N_c}} \int_0^L \underset{S.O.O.S.}{\cdots} \int_0^L$$

$$= \sum_{\substack{\text{all orderings} \\ \text{of species} \\ \text{with fixed } N_1,\ldots,N_c}} \prod_i N_i!$$

$$\times \int_0^L dX_N \underset{S.O.O.P.}{\cdots} \int_0^{X_2} dX_1 \qquad (9.4)$$

In the first step we have fixed a specific ordering of species (S.O.O.S.) and in the second step, we have fixed the specific ordering of the particles (S.O.O.P.) and multiplied by the factor $\prod_i N_i!$. [Note that the integrand in (9.1) is symmetric with respect to interchanging particles of the *same* species].

Applying (9.4) to (9.1) we get

$$Q(T, L, N) = \frac{1}{\prod_i \Lambda_i^{N_i}} \sum_{\substack{\text{all orderings} \\ \text{of species} \\ \text{with fixed } N_i, \ldots, N_c}}$$

$$\times \int_0^L dX_N \underset{S.O.O.P}{\cdots} \int_0^{X_2} dX_1 \exp[-\beta U(N)]$$

(9.5)

Note that in Eq. (9.5) $N_1, \ldots, N_c$ are fixed. Therefore, before taking the Laplace transform, we first open the system with respect to all species. The corresponding grand PF is

$$\Xi(T, L, \lambda) = \sum_N \prod_i \lambda_i^{N_i} Q(T, L, N) = \exp(\beta P L) \quad (9.6)$$

where $P = P(T, L, \lambda)$ is the thermodynamic pressure, expressed as a function of the variables T, L, λ. We now take the Laplace transform of (9.6) with respect to the variable L, by introducing the new unspecified variable P^*.

$$\Psi(\Xi) = \int_0^\infty \exp(-\beta P^* L) \Xi(T, L, \lambda) dL \quad (9.7)$$

As we have done in Section 8.1, we can identify the first particle with the Nth particle, and "close" the 1D system into a circle. This has no effect on the thermodynamics of the system.

Applying the convolution theorem we obtain the Laplace transform of Ξ,

$$\Psi(\Xi) = \int_0^\infty \exp(-\beta P^* L) \sum_N \prod_{N_i} \left(\frac{\lambda_i}{\Lambda_i}\right)^{N_i}$$

$$\times \sum_{\substack{all\ orderings \\ of\ species \\ with\ fixed\ N_1,\dots,N_c}} \int \overset{\cdots}{_{S.O.O.P}} \int$$

$$= \sum_{N=0}^\infty \sum_s M_{S_1 S_2} M_{S_2 S_3} \cdots M_{S_N S_1}$$

$$= \sum_{N=0}^\infty Tr M^N \qquad (9.8)$$

where the matrix elements are defined for each pair of species $\alpha\beta$ by

$$M_{\alpha\beta}(P^*) = (\lambda_\alpha/\Lambda_\alpha)^{1/2}(\lambda_\beta/\Lambda_\beta)^{1/2}\Psi_{\alpha\beta}(P^*) \qquad (9.9)$$

and $\Psi_{\alpha\beta}$ is defined by

$$\Psi_{\alpha\beta}(P^*) = \int_0^\infty \exp[-\beta P^* R - \beta U_{\alpha\beta}(R)]\, dR \qquad (9.10)$$

In (9.10), $\Psi_{\alpha\beta}$ is the Laplace transform of the function $\exp[-\beta U_{\alpha\beta}(R)]$. This is the generalization of (8.15) but now we choose the arbitrary variable P^* instead of the thermodynamic variable P.

The last step in (9.8) needs further elaboration. It is easier to visualize for a two-component system, for which we have the identity

$$\sum_{N_A=0}^{\infty} \sum_{N_B=0}^{\infty} \sum_{\substack{\text{all orderings} \\ \text{of species} \\ \text{with fixed } N_A, N_B}}$$

$$= \sum_{N=0}^{\infty} \sum_{N_A+N_B=N} \sum_{\substack{\text{all orderings} \\ \text{of species} \\ \text{with fixed } N_A, N_B}}$$

$$= \sum_{N=0}^{\infty} \sum_{s} \tag{9.11}$$

Note that the sum over "all orderings of species" is restricted to fixed values of N_A and N_B. Once we *open* the system, we carry the sum over N_A and N_B in two steps. First, we fix $N = N_A + N_B$. Next, the sum over all N_A and N_B with a fixed N, this is the same as the sum over all possible vectors s. This is followed by the sum over all N, which is the result on the rhs of (9.8).

Assuming that the sum of (9.8) converges, we rewrite it as

$$\Psi(\Xi) = \sum_{N=0}^{\infty} Tr M^N = \sum_{N=0}^{\infty} \sum_{j=1}^{c} \gamma_j^N = \sum_{N=0}^{\infty} [1 - \gamma_j(P^*)]^{-1} \tag{9.12}$$

where γ_j is the jth eigenvalue of the matrix. On the M rhs of Eq. (9.12), we stressed the dependence of γ_j on the variable P^*.

From the definition of Ξ and of $\Psi(\Xi)$ in (9.6) and (9.7), we obtain

$$\Psi(\Xi) = \int_0^\infty \exp[-\beta L(P^* - P)]dL \qquad (9.13)$$

If we choose P^* as the thermodynamic pressure P, then $\Psi(\Xi)$ is referred to as the generalized PF. It is clear that the integral (9.13) diverges. The reason is that $\Xi(T, L, \lambda)$ is a function of the single extensive variable L. Transforming the extensive variable L into the thermodynamic intensive variable P gives a partition function, which is a function of the intensive variables T, P, λ only. However, the Gibbs–Duhem relation states that

$$S\, dT - V\, dP + \sum N_i d\mu_i = 0 \qquad (9.14)$$

This means that the intensive variables T, L, μ or T, P, λ are not independent. For this reason, we have denoted by P^* the new variable in the Laplace transform taken in (9.7). Suppose that we choose $P^* > P$; then the integral in (9.13) converges and we have

$$\Psi(\Xi) = \frac{1}{\beta(P^* - P)} = \sum_{j=1}^c [1 - \gamma_j(P^*)]^{-1} \qquad (9.15)$$

Clearly, in the limit $P^* \to P$ the expression in (9.15) diverges. Therefore, there must be at least one eigenvalue, $\gamma_j(P^*)$ equal to 1.

In order to obtain the eigenvalues of M we need to solve the secular equation

$$|M - \gamma I| = 0 \qquad (9.16)$$

Recall that the elements of M are functions of T, P and λ. Therefore, we can use the implicit Eq. (9.16) to derive all thermodynamic quantities of interest.

From now we will discuss systems of two components A and B, for which the secular equation is

$$\begin{vmatrix} M_{AA} - \gamma & M_{AB} \\ M_{AB} & M_{BB} - \gamma \end{vmatrix} = 0 \qquad (9.17)$$

The two roots of this equation are

$$\gamma_\pm = \frac{1}{2}\left[M_{AA} + M_{BB} \pm \sqrt{(M_{AA} - M_{BB})^2 + 4M_{AB}^2} \right] \qquad (9.18)$$

The correct solution can be identified by taking the limit $\lambda_B = 0$, i.e., for pure A we must have

$$\gamma = \frac{1}{2}(M_{AA} + M_{AA}) = M_{AA} \qquad (9.19)$$

Therefore, we take the solution γ_+ in (9.18) and equate it to unity, to obtain the implicit equation of state:

$$f(T, P, \lambda_A, \lambda_B) = M_{AA} + M_{BB}$$

$$+ \sqrt{(M_{AA} - M_{BB})^2 + 4M_{AB}^2} - 2 = 0 \qquad (9.20)$$

Taking the total differential of f, we have

$$df = \frac{\partial f}{\partial T}dT + \frac{\partial f}{\partial P}dP + \sum \frac{\partial f}{\partial \lambda_i}d\lambda_i = 0 \qquad (9.21)$$

Comparing (9.21) with (9.14) we can derive all the thermodynamic quantities of the system from the implicit Eq. (9.20).

For example, the average density of A is

$$\rho_A = \left(\frac{\partial P}{\partial \mu_A}\right)_{T,\mu_B} = \frac{-\left(\partial f/\partial \mu_A\right)_{P,T,\mu_B}}{\left(\partial f/\partial P\right)_{T,\mu_A,\mu_B}} \qquad (9.22)$$

where the two derivatives on the rhs of (9.22) may be obtained from (9.20)

The entropy of the system is

$$S = \left(\frac{\left(\partial(PV)\right)}{\partial T}\right)_{V,\mu_A,\mu_B} = V\left(\frac{\partial P}{\partial T}\right)_{\mu_A,\mu_B}$$

$$= -V\left(\frac{\left(\partial f/\partial T\right)_{P,\mu_A,\mu_B}}{\left(\partial f/\partial P\right)_{T,\mu_A,\mu_B}}\right) \qquad (9.23)$$

To obtain the expression of the chemical potential, of say A, we can first eliminate λ_B from the equation of state (9.20), then substitute λ_B in the expression for the density in (9.22) to obtain ρ_A as a function of T, P, λ_A. This may be inverted to obtain $\lambda_A = \lambda_A(T, P, \rho_A)$ from which we get the expression for the chemical potential.

Before calculating the properties of the mixture, it is of interest to note a particularly simple case. If

$$\Psi_{AB}^2 = \Psi_{AA}\Psi_{BB} \qquad (9.24)$$

or equivalently

$$M_{AB}^2 = M_{AA}M_{BB} \qquad (9.25)$$

In this case the equation of state (9.20) reduces to

$$f(T, P, \lambda_A, \lambda_B) = 2(M_{AA} + M_{BB}) - 2 = 0 \qquad (9.26)$$

The density of A is obtained from (9.22)

$$\rho_A = \frac{-\beta \lambda_A}{\Lambda_A} \frac{2\Psi_{AA}}{\partial f / \partial P} \qquad (9.27)$$

and the mole fraction of A is given by

$$x_A = \frac{\rho_A}{\rho_A + \rho_B} = \frac{\lambda \Psi_{AA}}{\Lambda_A} \qquad (9.28)$$

This may be written in a more familiar form as (note that $\mu_A = k_B T \ln \lambda_A$)

$$\mu_A = k_B T \ln (\Lambda_A / \Psi_{AA}) + k_B T \ln x_A$$
$$= \mu_A^P + kT \ln x_A \qquad (9.29)$$

where μ_A^P is the chemical potential of pure A at the same T, P as of the mixture, see (8.21). Equation (9.29) is the familiar form of the chemical potential of A in a symmetrical ideal solution of A and B.[1]

9.2 The Global and Local Properties of Mixtures

Traditionally, the properties of mixtures were characterized by their excess function. These are *excesses* with respect to symmetrical ideal (*SI*) solutions. The excess quantities, defined below, provide *global* information on the mixture. Here, we used the term *global* only to contrast it with the term *local.* The local properties though equivalent and derivable from the global properties offer a host of new information on the *local* environments of each molecular species in the mixture.

In this section, we shall define the global quantities and outline the procedure for their calculation. We first need the expressions for the chemical potentials of A and B in the mixture. As noted in Section 9.1, we can obtain the density, say ρ_A from the equation of state. We first eliminate λ_B from the equation of state of (9.20), the result is

$$\lambda_B = \frac{\Lambda_B(\Lambda_A - \lambda_A \Psi_{AA})}{\lambda_A \Psi_{AB}^2 + (\Lambda_A - \lambda_A \Psi_{AA})\Psi_{BB}} \qquad (9.30)$$

This can be substituted on the rhs of (9.22) to obtain the density ρ_A as a function of T, P, λ_A. The latter can then be inverted to obtain $\lambda_A = \lambda_A(T, P, \rho_A)$. The explicit equation for λ_A as a function of T, P, ρ_A is quite complicated. However, for calculating the excess Gibbs energy, we do not need to calculate the pressure derivative in (9.22). The mole fraction x_A is calculated from

$$x_A = \frac{\rho_A}{\rho_A + \rho_B} \qquad (9.31)$$

having calculated $\rho_A(T, P, \lambda_A)$ and $\rho_B(T, P, \lambda_A)$, we can get $x_A(T, P, \lambda_A)$ from (9.31) and (9.22) without evaluating the denominator in (9.22). The result is:

$$x_A = \frac{\lambda_A \Lambda_A \Psi_{AB}^2}{\lambda_A(2\Lambda_A - \lambda_A \Psi_{AA})\Psi_{AB}^2 + (\Lambda_A - \lambda_A \Psi_{AA})^2 \Psi_{BB}} \qquad (9.32)$$

The last equation can be solved to obtain the absolute activity λ_A as a function of T, P, x_A. The result is

$$\lambda_A = \frac{\Lambda_A[(1 - 2x_A)\Psi_{AB}^2 + 2x_A\Psi_{AA}\Psi_{BB} \pm \Psi_{AB}\Gamma]}{2x_A\Psi_{AA}(\Psi_{AA}\Psi_{BB} - \Psi_{AB}^2)}$$

(9.33)

where we have denoted

$$\Gamma = \sqrt{(1 - x_A)^2\Psi_{AB}^2 + 4x_A(1 - x_A)\Psi_{AA}\Psi_{BB}} \qquad (9.34)$$

To determine the correct solution, we take the two limits: at $x_A \to 0$ and $x_A = 1$, we find that the "+" solution gives the correct limiting quantities. Thus, for $x_A \to 0$ we have

$$\lambda_A = \frac{\Lambda_A\Psi_{BB}x_A}{\Psi_{AB}^2} + O(x_A)^2 \qquad (9.35)$$

and for $x_A = 1$, we get the absolute activity of the pure A

$$\lambda_A^P = \frac{\Lambda_A}{\Psi_{AA}} \qquad (9.36)$$

Thus, from the "+" solution in (9.33), we can obtain the chemical potential of A. Similarly, we obtain the chemical potential of B. The excess Gibbs energy per particle as a function of (T, P, x_A) is defined by

$$g^{EX} = \frac{G^{EX}}{N_A + N_B} = x_A(\mu_A - \mu_A^P - k_BT \ln x_A)$$

$$+ (1 - x_A)(\mu_B - \mu_B^P - k_BT \ln(1 - x_A)) \qquad (9.37)$$

Once we have g^{EX} as a function of P, T, x_A we can get all the other excess thermodynamic quantities. For instance, the excess entropy and the excess volume may be calculated from

$$S^{EX} = - \left(\frac{\partial G^{EX}}{\partial T} \right)_{P,x_A} \tag{9.38}$$

$$V^{EX} = \left(\frac{\partial G^{EX}}{\partial P} \right)_{T,x_A} \tag{9.39}$$

The quantities G^{EX}, S^{EX}, V^{EX}, etc., are referred to as *global* quantities of the mixture. Note that these are excess quantities with respect to symmetrical ideal solutions. These should be distinguished from excess quantities with respect to dilute ideal solutions.[2]

To compute the *local* quantities, we also need the partial molar volumes and the isothermal compressibility of the mixture which are calculated from

$$V = \frac{\partial G}{\partial P} = \frac{\partial (N_A \mu_A + N_B \mu_B)}{\partial P} \tag{9.40}$$

$$\bar{V}_A = \left(\frac{\partial V}{\partial N_A} \right)_{T,P,N_B} \tag{9.41}$$

and

$$\kappa_T = - \frac{1}{V} \frac{\partial V}{\partial P} \tag{9.42}$$

Note that all the quantities g^{EX}, $\bar{V}_A$ and κ_T may be determined experimentally. One can show that these quantities are sufficient in determining the *local* quantities of interest through the inversion of the Kirkwood–Buff theory.[3]

The basic ingredients in the local characterization of the mixture are the Kirkwood–Buff integrals (*KBI*). These are defined for the 3D case as

$$G_{\alpha\beta} = \int_V [g_{\alpha\beta}(R) - 1]d\mathbf{R} = \int_0^\infty [g_{\alpha\beta}(R) - 1]4\pi R^2 dR$$

$$(9.43)$$

where $g_{\alpha\beta}$ is the (angular averaged) pair correlation function for the species α, β.[3] The integration in (9.43) is extended over the entire macroscopic volume V of the system. It should be stressed that these functions are defined in the open system with respect to both A and B. In a closed system, the normalization of the pair correlation function is

$$G_{\alpha\beta}^{closed} = -\delta_{\alpha\beta}/\rho_\alpha \qquad (9.44)$$

where ρ_α is the number density of the species α. The significance of $G_{\alpha\beta}$ in (9.43) as a local property follows from the following considerations: $\rho_\alpha g_{\alpha\beta}(R)4\pi R^2 dR$ is the average number of α particles in a spherical shell of radius R and width dR, centered by a β particle. $\rho_\alpha 4\pi R^2 dR$ is the average number of α particles in the same volume element $4\pi R^2 dR$ centered at a random point in the system, see Fig. 9.3. Thus, the integral

$$\rho_\alpha G_{\alpha\beta}(R_C) = \rho_\alpha \int_0^{R_C} [g_{\alpha\beta}(R) - 1]4\pi R^2 dR \qquad (9.45)$$

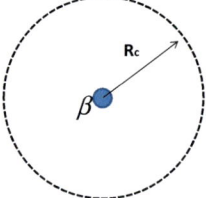

Fig. 9.3 A β particle at the center of a spherical shell of width dR and radius R.

measures the *change* in the average number of α particles in a spherical region of radius R_C caused by placing a β particle at the center of this region. The significance of $G_{\alpha\beta}(R_C)$ as a measure of the *local* properties around a β particle follows from the characteristic behavior of the pair correlation function. For most systems of interest (excluding solid solutions or systems near the critical point), $g_{\alpha\beta}(R)$ approaches unity (i.e. no correlation) at distances of a few molecular diameters. Hence, one can replace R_C in (9.45) by infinity. Thus, the major contribution to $G_{\alpha\beta}$ comes from a relatively small *local* region around the particle β. A positive value of $\rho_\alpha G_{\alpha\beta}$ means that when a β particle is placed at the center of the correlation volume (i.e. the volume $4\pi R_C^3/3$), it will attract an excess of α particles, compared to the average number of particles in the same spherical volume, at a randomly chosen point in the system. Because of this property and because $G_{\alpha\beta}$ is defined symmetrically with respect to α and β, i.e. $G_{\alpha\beta} = G_{\beta\alpha}$, we may refer to $G_{\alpha\beta}$ as a measure of the extent of *affinity* between the species α and β.

In the one dimensional case the *KBI* are defined by

$$G_{\alpha\beta} = 2 \int_0^\infty [g_{\alpha\beta}(R) - 1] \, dR = \int_{-\infty}^\infty [g_{\alpha\beta}(R) - 1] \, dR$$

(9.46)

where the integration over the entire "volume" of the system can be carried out either as twice the integration for positive values of R, or integrate over the two sides of the center of the particle, from minus to plus infinity.

The second *local* quantity of interest is the preferential solvation (*PS*). In a two-component system of A's and B's, we focus again on a small region V_C and ask how the *composition* in this region is altered by placing, say an A particle in its center. We define the *PS* of A with respect to composition, as the difference

$$\delta_{A,A}(V_C) = x_A^L(A) - x_A \qquad (9.47)$$

where x_A is the *bulk* mole fraction of A, and in x_A^L is the *local* mole fraction of A in the volume V_C around an A particle. From the definition of the *KBI*'s, it follows that

$$\delta_{A,A}(V_C) = \frac{x_A x_B (G_{AA} - G_{AB})}{x_A G_{AA} + x_B G_{AB} + V_C} \qquad (9.48)$$

Here, $G_{\alpha\beta}$ are the *KBI*'s as defined in (9.43). Therefore, (9.47) is valid for V_C *larger* than the correlation volume.

Similarly, the *PS* of B with respect to the composition (defined in terms of the mole fraction of A) is defined as

$$\delta_{A,B}(V_C) = x_A^L(B) - x_A$$

$$= \frac{x_A x_B (G_{AB} - G_{BB})}{x_A G_{AB} + x_B G_{BB} + V_C} \qquad (9.49)$$

Clearly, these quantities depend on the choice of the correlation volume V_C. For very large V_C, the local composition should approach the bulk composition, hence the *PS* should approach zero. In order to eliminate the dependence on V_C and obtain an *intrinsic* measure *PS* of A and B, we take the first term in the expansion of $\delta_{A,B}(V_C)$ in power series around V_C^{-1}. The coefficients of the first order terms in the expansion are

$$\delta_{A,A}^0 = x_A x_B (G_{AA} - G_{AB}) \qquad (9.50)$$

$$\delta_{A,B}^0 = x_A x_B (G_{AB} - G_{BB}) \qquad (9.51)$$

It should be noted that in expanding the quantities $\delta_{A,A}$ and $\delta_{A,B}$, we do *not* view $G_{\alpha\beta}$ as being functions of V_C. The definition (9.43) or (9.45), i.e. with $R_C \to \infty$, renders all $G_{\alpha\beta}$ independent of V_C. The only dependence on V_C is through the denominator in (9.48) and (9.49).

The third local quantity, which can be obtained from the *KBI*, measures deviations from *SI* solution. This quantity is defined by

$$\Delta_{AB} = G_{AA} + G_{BB} - 2G_{AB} \qquad (9.52)$$

This quantity depends on all three *KBI* (note that $G_{AB} = G_{BA}$).

The Kirkwood–Buff theory provides an exact relationship between the composition derivative of the chemical potential, and Δ_{AB}. We quote here the result.[2]

$$\left(\frac{\partial \mu_A}{\partial x_A} \right)_{P,T} = k_B T \left(\frac{1}{x_A} - \frac{x_B \rho_T \Delta_{AB}}{1 + \rho_T x_A x_B \Delta_{AB}} \right) \qquad (9.53)$$

We define a symmetrical ideal (*SI*) solution of two-components A and B, whenever the chemical potential, of say, A has the form

$$\mu_A = \mu_A^P + k_B T \ln x_A \quad \text{for } 0 \le x_A \le 1 \qquad (9.54)$$

where μ_A^P is the chemical potential of pure A at the same T and P as those of the mixture. From Eq. (9.53) and from the

Gibbs–Duhem relationship, it follows that

$$\mu_B = \mu_B^P + k_B T \ln x_B \quad \text{for } 0 \le x_B \le 1 \qquad (9.55)$$

It follows from (9.53) that $\Delta_{AB} = 0$ is a necessary and sufficient condition for SI behavior. A finite value of Δ_{AB} measures the extent of the deviation from SI behavior. More precisely, in the general case we write:

$$\mu_A = \mu_A^P + k_B T \ln x_A + k_B T \ln \gamma_A^{SI}$$
$$= \mu_A^{SI} + \mu_A^{EX} \qquad (9.56)$$

where

$$\mu_A^{EX} = k_B T \ln \gamma_A^{SI} = k_B T \int_0^{X_B} \frac{\rho T x_B' \Delta_{AB}}{1 + \rho T x_A' x_B' \Delta_{AB}} dx_B' \qquad (9.57)$$

where $x_B = 1 - x_A$ and Δ_{AB} is viewed as a function of x_B'. Thus, knowing $\Delta_{AB}(x_B)$, one can estimate the deviation from SI, either in terms of an activity coefficient or in terms of excess chemical potential.

Note that the condition $\Delta_{AB} = 0$ is equivalent to the conditions

$$\delta_{A,A}^0 = \delta_{A,B}^0 = -\delta_{B,B}^0 \qquad (9.58)$$

Hence, in a symmetrical ideal solution, the limiting preferential solvation, $\delta_{A,A}^0$ and $\delta_{A,B}^0$ have equal value but not necessarily zero. Thus, both Δ_{AB}, and the components $\delta_{A,A}^0$ and $\delta_{A,B}^0$ are important in analyzing the sources of non-ideality of the mixture.

The last *local* property is the solvation Gibbs energy of both A and B in the entire range of compositions $0 \leq x_A \leq 1$. Traditionally, the solvation properties of a molecule A in a solvent B were defined only in the limit of very dilute solutions. With the new definition of the solvation process the solvation properties of A can be defined and measured in the entire range of composition, including the solvation of A in pure A.[2]

The solvation Gibbs energy may be obtained from the pseudo-chemical potential as follows. For any mixture of A and B we write the chemical potential of say, A as

$$\mu_A = \mu_A^* + k_B T \ln \rho_A \Lambda_A \qquad (9.59)$$

The expression for the chemical potential may be obtained from the equation of state (9.20)

$$\rho_A(T, P, \lambda_A, \lambda_B) = \frac{-\partial f / \partial \mu_A}{\partial f / \partial P} \qquad (9.60)$$

by substituting λ_B from the equation of state (9.20) and inverting (9.60) we can obtain λ_A as a function of T, P, ρ_A. The expression is quite complicated. However, for a system of A very dilute B, i.e., when $\lambda_A \ll \lambda_B$ we can get a simple expression.

Expanding (9.18) to first order in λ_A, we obtain for the density of A, noting that for pure B we have $\lambda_B \Psi_{BB} / \Lambda_B = 1$, hence the density of A in this limit

$$\rho_A = \left(\frac{\partial P}{\partial \mu_A} \right)_T = -\frac{\partial f / \partial \mu_A}{\partial f / \partial P} = -\beta \frac{\lambda_A}{\Lambda_A} \frac{\Psi_{AB}^2}{\Psi_{BB}'} \qquad (9.61)$$

where Ψ'_{BB} is the derivative of Ψ_{BB} with respect to P. The last equation can be arranged to obtain

$$\mu_A = k_B T \ln \rho_A \Lambda_A - k_B T \ln \frac{-\beta \Psi^2_{AB}}{\Psi'_{BB}} \qquad (9.62)$$

From (9.62) we identify the pseudo-chemical potential

$$\mu^*_A = -k_B T \ln \frac{-\beta \Psi^2_{AB}}{\Psi'_{BB}} \qquad (9.63)$$

which is the generalization of (8.52). Having $\mu^*_A(P, T)$, we can obtain all other thermodynamic quantities of solvation of A in B.

Clearly, if we have the correlation functions $g_{\alpha\beta}(R)$ we can calculate all the KBIs. In practice, we can calculate all the KBI from experimental quantities as follows:

First, we can calculate all the KBI from the inversion of the Kirkwood–Buff theory. The results are[2]

$$G_{AA} = k_B T \kappa_T - \frac{1}{\rho_A} + \frac{\rho_B \bar{V}^2_B}{x_A D} \qquad (9.64)$$

$$G_{BB} = k_B T \kappa_T - \frac{1}{\rho_B} + \frac{\rho_A \bar{V}^2_A}{x_B D} \qquad (9.65)$$

$$G_{AB} = k_B T \kappa_T - \frac{\bar{V}_A \bar{V}_B (\rho_A + \rho_B)}{D} \qquad (9.66)$$

where $\rho_A + \rho_B = \rho_T$, and

$$D = 1 + x_A x_B \left(\frac{\partial^2 \left(g^{EX}/k_B T \right)}{\partial x^2_A} \right)_{P,T} \qquad (9.67)$$

Note that if the system is an ideal gas, then by definition of an ideal gas (in the sense that all $U_{ij}(R) \equiv 0$, not only in the limit $\rho_T \to 0$) all $G_{ij} = 0$.

The extent of deviation from SI solution, $\rho_T \Delta_{AB}$ can be calculated either from (9.68) or from the excess Gibbs energy per molecule of mixture $g^{EX} = G^{EX}/(N_A + N_B)$ and the relationship

$$\left(\frac{\partial^2 g}{\partial x_A^2} \right)_{P,T} = \frac{k_B T}{x_A x_B (1 + x_A x_B \rho_T \Delta_{AB})} \tag{9.68}$$

where

$$\Delta_{AB} = G_{AA} + G_{AB} - 2G_{AB} \tag{9.69}$$

Finally, the solvation Gibbs energy of say A relative to the solvation Gibbs energy of A in pure A is defined by

$$\Delta\Delta\mu_A^* = \Delta\mu_A^*(in\ the\ mixture) - \Delta\mu_A^*(in\ pure\ A) \tag{9.70}$$

This quantity may also be calculated from the excess Gibbs energy and excess volume by

$$\Delta\Delta\mu_A^* = k_B T \ln \left[\frac{V^{EX} + x_A V_A^P + x_B V_B^P}{V_A^P} \right]$$

$$+ g^{EX} + x_B \left(\frac{\partial g^{EX}}{\partial x_A} \right)_{P,T} \tag{9.71}$$

9.3 (***) Mixtures of Hard-Rods

We start with the simplest systems consisting of mixtures of hard rods (HR). The HR are defined in terms of the pair potential

$$U(R) = \begin{cases} \infty & \text{for } R \leq \sigma \\ 0 & \text{for } R > \sigma \end{cases} \tag{9.72}$$

where σ is the "diameter", or the length of the rods.

In these systems, the only molecular parameters that can be varied is the ratio of the diameters of the particles. (For a complete characterization of the system, the masses of the particles should also be specified. However, the masses do not affect any of the properties we are interested in, in this section).

For *HRs*, the functions $\Psi_{\alpha\beta}$ reduces to

$$\Psi_{\alpha\beta}(T, P) = \int_0^\infty \exp[-\beta P R - \beta U_{\alpha\beta}(R)]\, dR$$

$$= \exp[-\beta P \sigma_{\alpha\beta}]/\beta P \tag{9.73}$$

For *HRs* we also have

$$\sigma_{AB} = \frac{\sigma_{AA} + \sigma_{BB}}{2} \tag{9.74}$$

Therefore, it follows that

$$\Psi_{AB}^2 = \Psi_{AA}\Psi_{BB} \tag{9.75}$$

Hence, a mixture of *HRs* *always* forms a symmetric ideal (*SI*) solution (see Eqs. (9.24) to (9.29)). Hence, for such mixtures all the excess thermodynamic quantities are zeros. Note that this result holds true only for a 1D system. It is not true for mixtures of hard spheres in 3D.

The chemical potentials of A and B are given by

$$\mu_A(T, P, x_A) = P\sigma_{AA} + k_B T \ln(\beta P \Lambda_A x_A) \qquad (9.76)$$

$$\mu_B(T, P, x_A) = P\sigma_{BB} + k_B T \ln[\beta P \Lambda_B(1 - x_A)] \qquad (9.77)$$

The partial molar volumes of the two components are

$$\bar{V}_A = \sigma_{AA} + (\beta P)^{-1}, \quad \bar{V}_B = \sigma_{BB} + (\beta P)^{-1} \qquad (9.78)$$

The volume, per particle of the mixture, is

$$v = \frac{V}{N_A + N_B} = x_A \bar{V}_A + (1 - x_A) \bar{V}_B$$

$$= x_A \sigma_{AA} + (1 - x_B) \sigma_{BB} + (\beta P)^{-1} \qquad (9.79)$$

Figure 9.4A shows the volume per particle of a mixture of HRs with diameters $\sigma_{AA} = 1, \sigma_{BB} = 2, T = 1$. For different

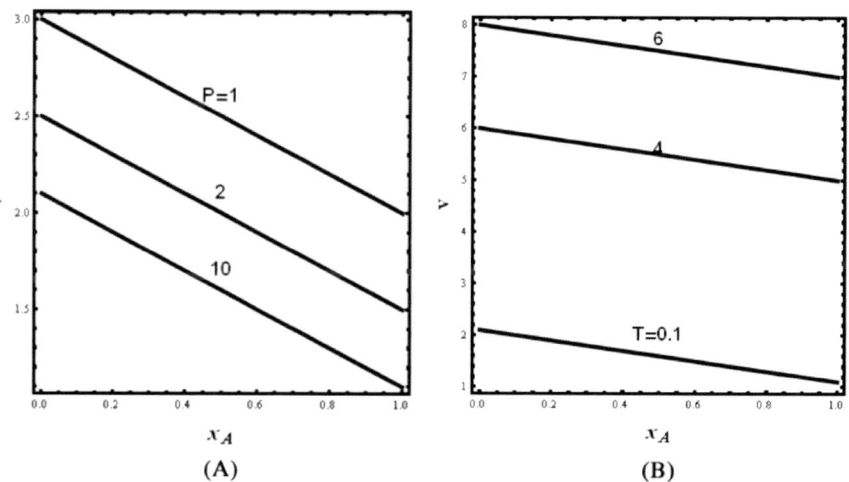

(A)   (B)

Fig. 9.4 The volume of the mixture A and B as a function of the mole fraction of A, for hard rods with $\sigma_{AA} = 1$ and $\sigma_{BB} = 2$.

pressures as the pressure increases, or as the temperature decreases, the curves converge to the limiting linear curve

$$v = x_A \sigma_{AA} + (1 - x_A) \sigma_{BB} \tag{9.80}$$

Figure 9.4B shows the volume per particle at $P = 1$ and different temperatures.

Figure 9.5 shows the volume as a function of pressure for different temperatures (and $x_A = 0.5$). Note that in the 1D system, there is no phase separation, and $\partial V / \partial P$ is always positive.

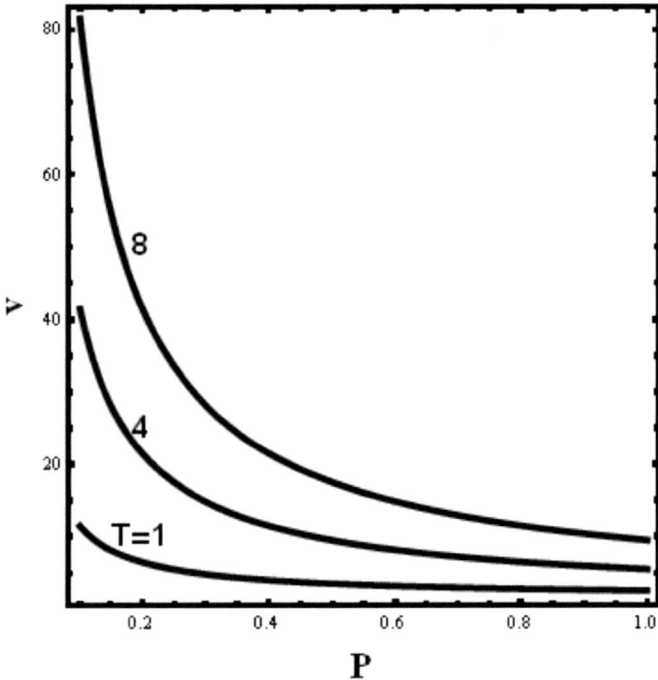

Fig. 9.5 The volume per particle as a function of P at different temperatures and $x_A = 0.5$.

The compressibility of the mixture is given by

$$\kappa_T = -\frac{1}{V}\left(\frac{\partial V}{\partial P}\right)_T$$

$$= \frac{1}{P[1 + \beta P(x_A \sigma_{AA} + (1 - x_A)\sigma_{BB})]} \tag{9.81}$$

Figure 9.6 shows the compressibility as a function of x_A, at a fixed temperature ($k_B T = 1$) and various pressures. As expected, the compressibility is large at low pressures, but at high pressures it converges to zero, for any composition of the system.

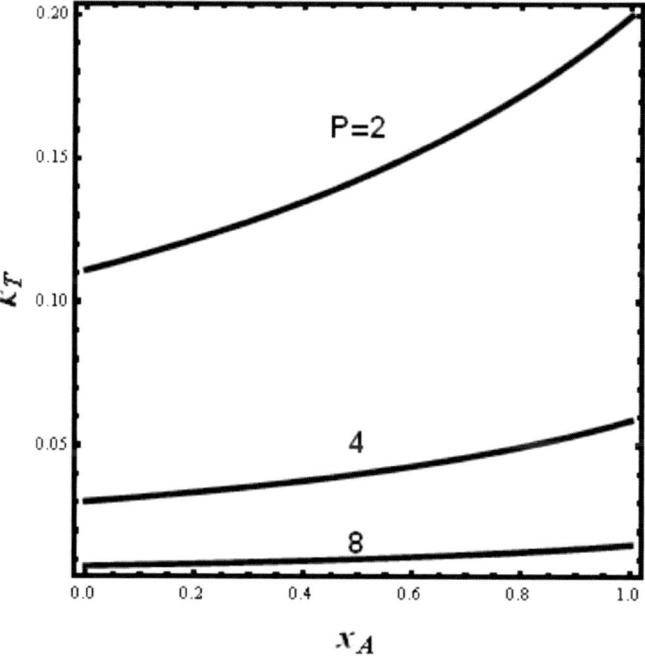

Fig. 9.6 The isothermal compressibility of the same system as in Fig. 9.4 as a function x_A of at a fixed $T = 1$ and different pressures.

Next, we calculate the *KB* integrals. These were calculated from Eqs. (9.64) to (9.67), which for *HR* reduces to

$$G_{AA} = \rho_T \langle \sigma^2 \rangle - 2\sigma_{AA}$$
$$G_{BB} = \rho_T \langle \sigma^2 \rangle - 2\sigma_{BB}$$
$$G_{AB} = \rho_T \langle \sigma^2 \rangle - \sigma_{AA} - \sigma_{BB} \tag{9.82}$$

where $\rho_T = \rho_A + \rho_B$ is the total number density of the particles, and

$$\langle \sigma^2 \rangle = x_A \sigma_{AA}^2 + (1 - x_A)\sigma_{BB}^2 \tag{9.83}$$

In the limit, $\rho_B \to 0$ (or $x_A \to 1$). These quantities reduce to ($\rho_T \to \rho_A$)

$$G_{AA} = \rho_A \sigma_{AA}^2 - 2\sigma_{AA}$$
$$G_{BB} = \rho_A \sigma_{AA}^2 - 2\sigma_{BB}$$
$$G_{AB} = \rho_A \sigma_{AA}^2 - \sigma_{AA} - \sigma_{BB} \tag{9.84}$$

And similar equations for the limit $\rho_A \to 0$ (or $x_A \to 0$), i.e.,

$$G_{AA} = \rho_T \sigma_{BB}^2 - 2\sigma_{AA}$$
$$G_{BB} = \rho_T \sigma_{BB}^2 - 2\sigma_{BB}$$
$$G_{AB} = \rho_T \sigma_{BB}^2 - \sigma_{AA} - \sigma_{BB} \tag{9.85}$$

Two other limiting cases are of interest. At very low densities $\rho_T \to 0$, we have

$$G_{AA} = -2\sigma_{AA}$$
$$G_{BB} = -2\sigma_{BB}$$
$$G_{AB} = -\sigma_{AA} - \sigma_{BB} \tag{9.86}$$

It is known that at this limit, the pair correlation functions are given by[2]

$$g_{\alpha\beta}(R) \approx \exp[-\beta U_{\alpha\beta}(R)] \qquad (9.87)$$

hence, the *KB* integrals in (9.86) are all negative, and equal to twice the distance of closest approach. Clearly, this is due to the repulsive part of the hardcore potential function (9.72).

The other limit is very high density. There is a limit on the total density, which can be obtained in this system. The limiting density can be obtained from (9.79) either at $P \to \infty$ or at $T \to 0$, and is

$$\rho_{T,max} \to (x_A \sigma_{AA} + (1 - x_A)\sigma_{BB})^{-1} = \langle \sigma \rangle^{-1} \qquad (9.88)$$

at this limit, we have

$$G_{AA} = \frac{\langle \sigma^2 \rangle}{\langle \sigma \rangle} - 2\sigma_{AA}$$

$$G_{BB} = \frac{\langle \sigma^2 \rangle}{\langle \sigma \rangle} - 2\sigma_{BB}$$

$$G_{AB} = \frac{\langle \sigma^2 \rangle}{\langle \sigma \rangle} - \sigma_{AA} - \sigma_{BB} \qquad (9.89)$$

Figure 9.7 shows G_{AA}, G_{BB} and G_{AB} as a function of x_A for three total densities. Note that the maximum density of pure A is $\rho_{max,A} = 1$, and of pure $B \rho_{max,B} = \frac{1}{2}$. Therefore, we plotted the values of $G_{\alpha\beta}$ for densities below the maximal density of the mixture (9.88).

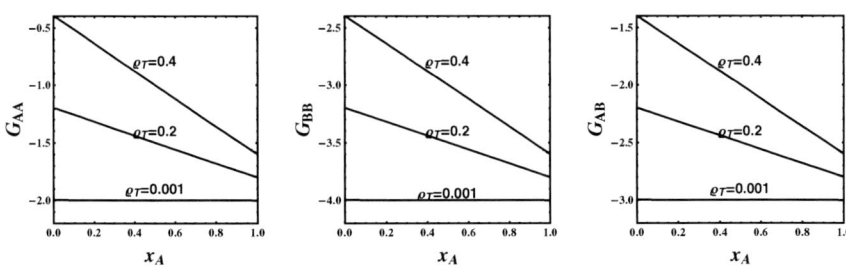

Fig. 9.7 Values of the *KBI*s for the same system as in Fig. 9.4, and at different total densities.

The limiting coefficients of the preferential solvation are

$$\delta_{A,A}^0 = x_A x_B (G_{AA} - G_{BB}) = (\sigma_{BB} - \sigma_{AA}) x_A x_B$$

$$\delta_{B,B}^0 = x_A x_B (G_{BB} - G_{AB}) = (\sigma_{AA} - \sigma_{BB}) x_A x_B \quad (9.90)$$

and the measure of the deviation from symmetric ideal solution is:

$$\Delta_{AB} = G_{AA} + G_{BB} - 2G_{AB} = (\sigma_{BB} - \sigma_{AA})$$

$$+ (\sigma_{AA} - \sigma_{BB}) = 0 \quad (9.91)$$

As we have seen a mixture of hard rods is a *SI* solution, i.e. $\Delta_{AB} = 0$. From (9.91), we see that the *SI* behavior arises from the cancellation of the two terms $\delta_{A,A}^0$ and $\delta_{B,B}^0$, which in general are non-zero, and in our case $\sigma_{BB} - \sigma_{AA} = 1$ and $\sigma_{AA} - \sigma_{BB} = -1$.

Finally, Figs. 9.8 and 9.9 show the solvation Gibbs energies of *A* and *B* at various temperatures and at various pressures. For the *HR*s mixtures, the solvation Gibbs energies are:

$$\Delta G_A^* = P\sigma_{AA} + k_B T \ln (1 + \beta P \langle \sigma \rangle) \quad (9.92)$$

$$\Delta G_B^* = P\sigma_{BB} + k_B T \ln(1 + \beta P \langle \sigma \rangle) \quad (9.93)$$

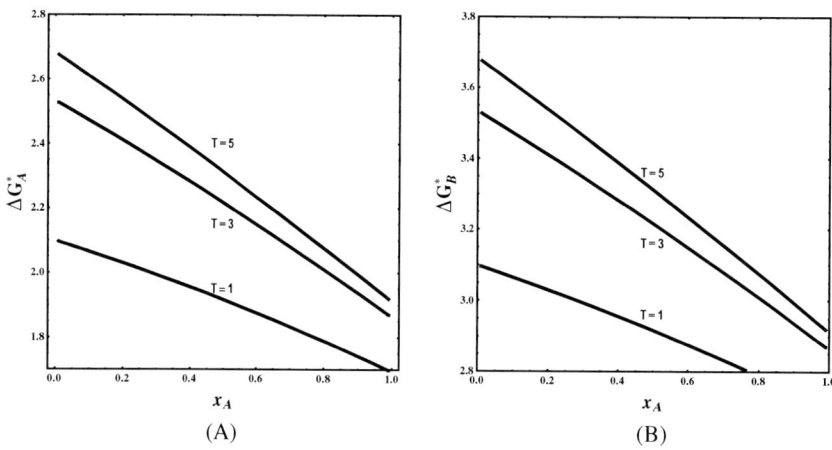

Fig. 9.8 The solvation Gibbs energy of A and B for the same system as in Fig. 9.4. at different temperatures.

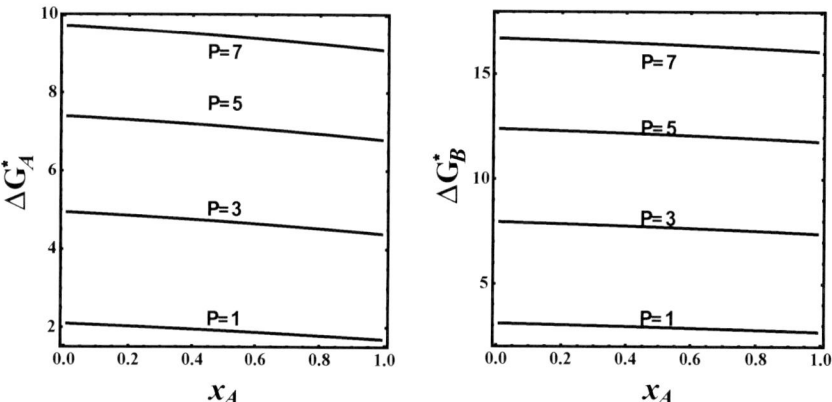

Fig. 9.9 The solvation Gibbs energy of A and B for the same system as in Fig. 9.4 at different pressures.

In the system of HR, the solvation Gibbs energy is related to the work of creating a cavity produced by the HR particles, which is "solvated" in the mixture. The larger the temperature, or the pressure, the greater the work required to create such a cavity.

9.4 (***) Mixture of Square-Well (SW) Particles

In this section, we present some results on the global and local properties of mixtures of particles interacting via a square-well potential, which we will refer to as *SW*-particles. These particles are defined by the pair potential

$$
U(R) = \begin{cases} \infty & \text{for } R < \sigma \\ \varepsilon & \text{for } \sigma \leq R < \sigma + \delta \\ 0 & \text{for } R > \sigma + \sigma \end{cases} \tag{9.94}
$$

We have seen that systems of hard rods from *SI* solutions. Therefore, all the excess thermodynamic quantities are zero. We have also examined the dependence of the local properties on the ratio of diameters in Section 9.3. Therefore, in this section, we choose equal diameters for the particles $\sigma_{AA} = \sigma_{BB} = 1$, and explore the dependence of the thermodynamic properties of the mixture on the ratio of the energy parameter ε. Before we move on to numerical examples it is instructive to note that in all the mixtures discussed in this section, the two components form a single stable phase with no phase separation.[3]

The Laplace transform as defined in (8.15) for the SW potential is:

$$
\Psi_{\alpha\beta}(P, T) = \int_0^\infty \exp[-\beta PR - \beta U_{\alpha\beta}(R)] \, dR
$$

$$
= \frac{\exp[-\beta P(\sigma_{\alpha\beta} + \delta)] + \exp[-\beta\varepsilon_{\alpha\beta} - \beta P\sigma_{\alpha\beta}](1 - \exp[-\beta\delta])}{\beta P}
$$

$$\tag{9.95}$$

For $\delta > 0$, $1 - \exp[-\beta\delta] > 1$, therefore the entire integral is positive.

The second derivative of the Gibbs energy per mole of the mixture $[g = G/(N_A + N_B)]$ is

$$\left(\frac{\partial^2 g}{\partial x^2}\right)_{T,P}$$

$$= \frac{k_B T \Psi_{AB}}{x(1-x)\sqrt{(1-2x)^2 \Psi_{AB}^2 + 4x(1-x)\Psi_{AA}\Psi_{BB}}}$$

$$(9.96)$$

Clearly, this is also positive in the entire range of compositions. Note also that since the square root in (9.96) tends to Ψ_{AB} at either $x \to 0$, or $x \to 1$ the entire second derivative diverges to infinity at these limits. Also note that at $x \to 0.5$ the second derivative is

$$\left(\frac{\partial^2 g}{\partial x^2}\right)_{T,P} = \frac{4k_B T \psi_{AB}}{\sqrt{\psi_{ABT}\psi_{BB}}} \qquad (9.97)$$

In the rest of this section we show a few results based on calculations with the following dimensionless parameters:

$$\sigma_{AA} = \sigma_{AB} = \sigma_{BB} = 1, \quad \varepsilon_{AA} = -1, \quad \varepsilon_{BB} = \varepsilon,$$

$$\varepsilon_{AB} = \sqrt{\varepsilon_{AA}\varepsilon_{BB}} \qquad (9.98)$$

and chose $T = 1$ and $P = 1$ for the numerical illustrations.

Figure 9.10 shows the volume per particle as a function of the mole fraction x_A for various values of the energy parameter ε. Since we have chosen equal diameters for the

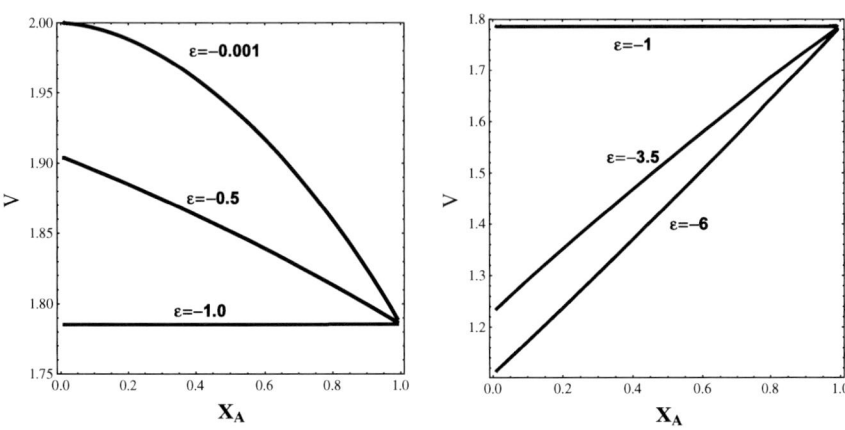

Fig. 9.10 The volume of a system of A and B particles interacting via square well potential with parameters as in Eq. (9.98) for different values of ε.

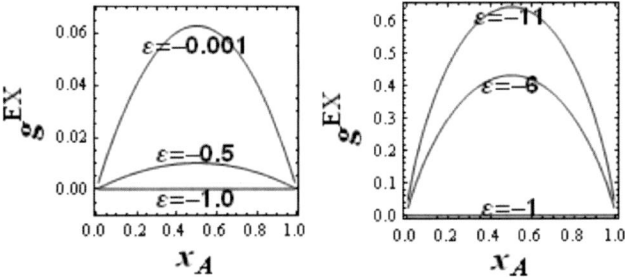

Fig. 9.11 The excess Gibbs energy for the same system as in Fig. 9.10.

particles, the case $\varepsilon = -1$ corresponds to a *SI* solution. In this illustration we keep T and P are constant, hence the volume of the system increases or decreases according to the corresponding increase or decrease of $|\varepsilon|$.

Figure 9.11 shows the excess Gibbs energy of the mixture for the same set of molecular parameters at $\varepsilon = -1$; we have a *SI* solution and $g^{EX} = 0$. As $|\varepsilon|$ either increases or decreases, we find positive deviations from *SI* behavior. Note that in

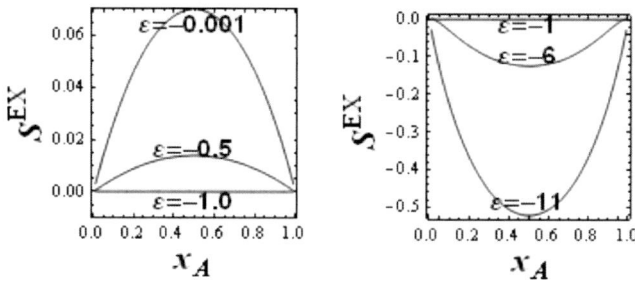

Fig. 9.12   The excess entropy for the same system as in Fig. 9.10.

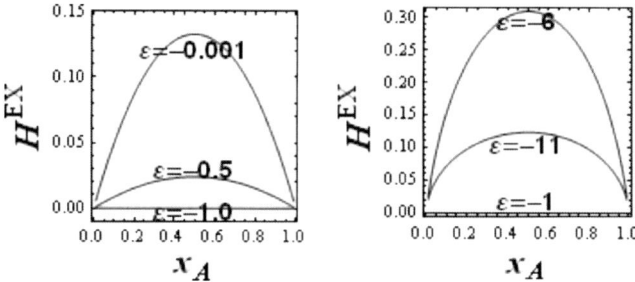

Fig. 9.13   The excess enthalpy for the same system as in Fig. 9.10.

this case there exists no phase separation. The Gibbs energy is always a concave downwards function of x_A.

Figure 9.12 shows the excess entropy of the system. Note that as $|\varepsilon|$ decreases, we get larger and positive s^{EX}. However, as $|\varepsilon|$ increases above $|\varepsilon| = 1$, initially s^{EX} is positive, and for $|\varepsilon| = 5$ and larger it becomes negative.

Figure 9.13 shows the excess enthalpies of the system for the same set of parameters. As can be seen from the figure, all values of the excess enthalpies are positive.

Figure 9.14 shows the excess volume for the same set of parameters. For $|\varepsilon| \leq 1$, v^{EX} is positive and decreases with $|\varepsilon|$. This makes sense; by weakening the overall interactions in the system, keeping T and P constants, the volume expands.

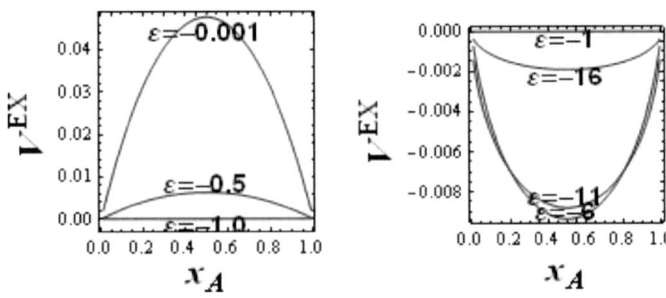

Fig. 9.14 The excess volume for the same system as in Fig. 9.10.

The smaller the $|\varepsilon|$, the larger the excess volume. On the other hand, for $|\varepsilon| \geq 1$ we find an initial increase in v^{EX} but as $|\varepsilon|$ becomes larger, the strong interaction causes an overall contraction of the system, and v^{EX} becomes negative.

We now turn to a few illustrations of the local quantities. We use the same set of molecular parameters, and fixed temperature $T = 1$, and pressure $P = 1$, as in the previous illustrations.

As a general comment, we note that when $|\varepsilon| = 1$, the system is a SI solution, and as we have seen, all the excess thermodynamic functions are zero. The local quantities, on the other hand are not zero.

The interpretation of the results is sometimes straight-forward, sometimes more difficult. In most cases, we can interpret the sign and the magnitude of the KBI by considering the split of the integral as follows

$$G_{ij} = \int_{-\infty}^{\infty} [g_{ij}(R) - 1]dR = -2\sigma_{ij} + A_{ij} \qquad (9.99)$$

where $-2\sigma_{ij}$ is a result of the direct repulsion between the two particles i and j, and A_{ij} is due to the correlation over

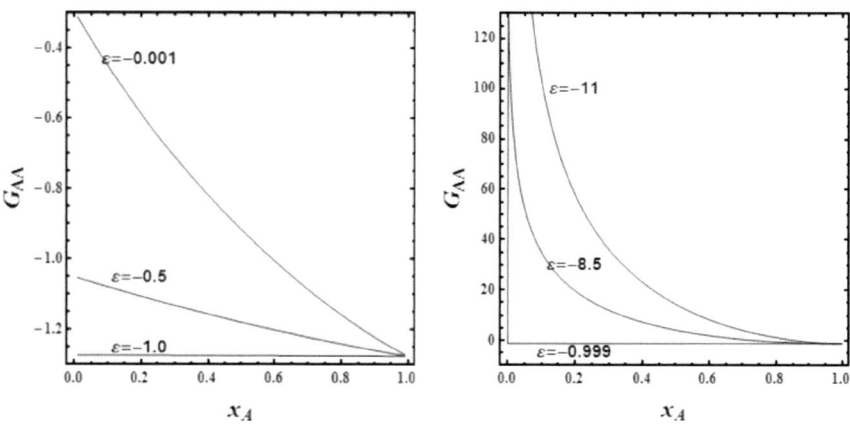

Fig. 9.15 The *KBI* G_{AA} for the same system as in Fig. 9.10.

distances beyond the hardcore diameters of the particles. The first term is always negative and is proportional to the size of the particles. The second, can be either positive or negative depending on the form of the pair correlation function in the region $\sigma_{ij} \leq R \leq \infty$.

Figure 9.15 shows G_{AA} as a function of the mole fraction of A. For small values of $|\varepsilon| \leq 1$, we see that G_{AA} is negative. This is mainly due to the repulsive part of the interaction, as we observed for *HR*s in Fig. 9.7. As we increase $|\varepsilon|$, the *KBI* becomes larger and less dependent on the composition. At $|\varepsilon| = 1$, the values of G_{AA} are independent of the composition and attain the value of G_{AA} for pure A. Note also that all the curves converge to the same value for $x_A = 1$, i.e., the value of G_{AA} for pure A, which in this case is about -1.22. Once we increase $|\varepsilon|$ beyond $|\varepsilon| > 1$, the affinities between pairs of A particles increase due to the stronger cohesive forces of the "solvent", which is the B component when $x_A \approx 0$. The larger the $|\varepsilon|$, the larger

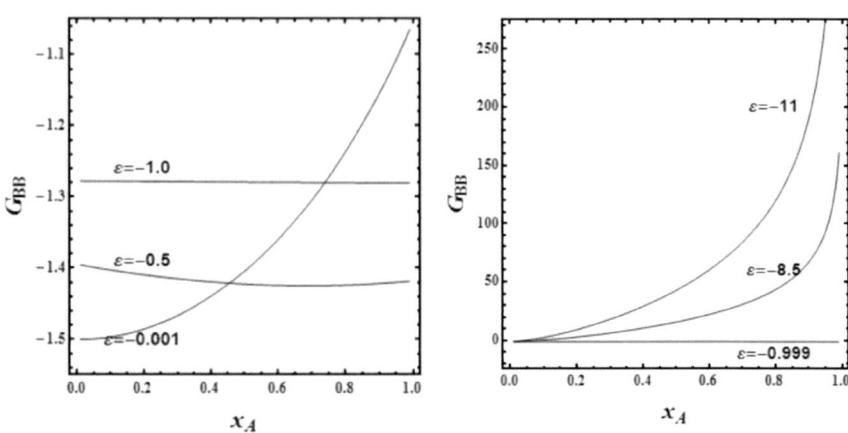

Fig. 9.16 The *KBI* G_{BB} for the same system as in Fig. 9.10.

and positive values of G_{AA}. This behavior is relevant to the phenomena of hydrophobic interactions which we shall not discuss here. Note again that all the curves converge to the limiting value of pure A at $x_A \to 1$.

Figure 9.16 shows the values of G_{BB} for the same set of parameters. Here, the behavior is quite different from that of G_{AA}. When $x_A = 0$, i.e., for pure B, the values of G_{BB} still changes as a function of ε. As we increase $|\varepsilon|$, the affinities between two B particles become less and less negative. Note that for $|\varepsilon| = 1$, G_{BB} is independent of x_A, but it is not zero. Again, we see that when $|\varepsilon|$ increases beyond one, the values of G_{BB} become large and positive.

It should be noted that the molecular reasons for the large positive values of G_{BB} at $x_A \approx 1$ are similar to the large positive values of G_{AA} at $x_A \approx 0$. In the case of G_{BB} at $x_A \approx 1$, we have B diluted in A. The stronger the BB interaction, the larger positive affinities between two B particles.

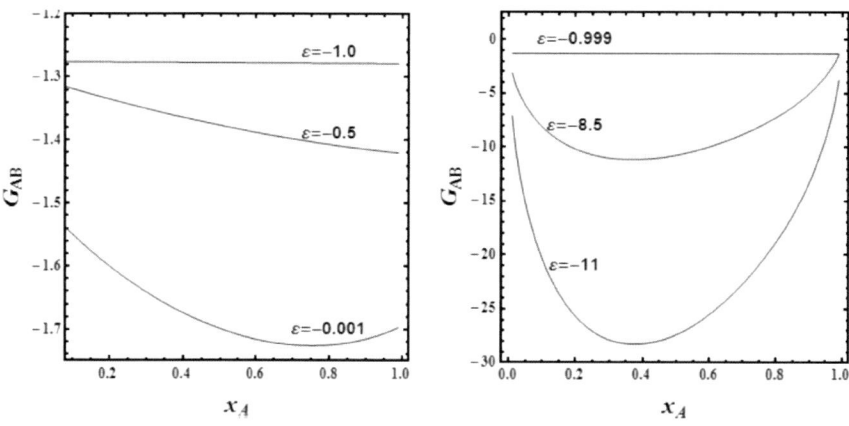

Fig. 9.17   The *KBI* G_{AB} for the same system as in Fig. 9.10.

Figure 9.17 shows the behavior of G_{AB}, which is quite different from the behavior of both G_{AA} and G_{BB}. The reason is that in neither $x_A = 0$, nor in $x_A \approx 1$, we have a *KBI* of "pure" *AB*. Here again, we find for $|\varepsilon| = 1$, a non-zero value of G_{AB}, but independent of x_A. For small values of $|\varepsilon| \leq 1$, G_{AB} is negative but decreases in absolute magnitude when $|\varepsilon|$ increases towards $|\varepsilon| = 1$. On the other hand, for $|\varepsilon| > 1$, we see that G_{AB} become large and negative as $|\varepsilon|$ increases.

Next, we turn to the quantity Δ_{AB}, Fig. 9.18, defined in Eq. (9.52) which is a measure of the deviations from *SI* solution. As expected, for $|\varepsilon| = 1$, we find that $\Delta_{AB} = 0$, i.e. the system forms a *SI* solution. For $|\varepsilon| \leq 1$, Δ_{AB} is larger the larger dissimilarity between the particles, i.e. the smaller the value of $|\varepsilon| \leq 1$. However, for $|\varepsilon| \geq 1$, we observe larger positive deviations from *SI*. The deviations seem to get very large at both ends of the composition range.

In Fig. 9.19, we present the limiting coefficient of the preferential solvation (*PS*) of *A* with respect to *A* and *B*.

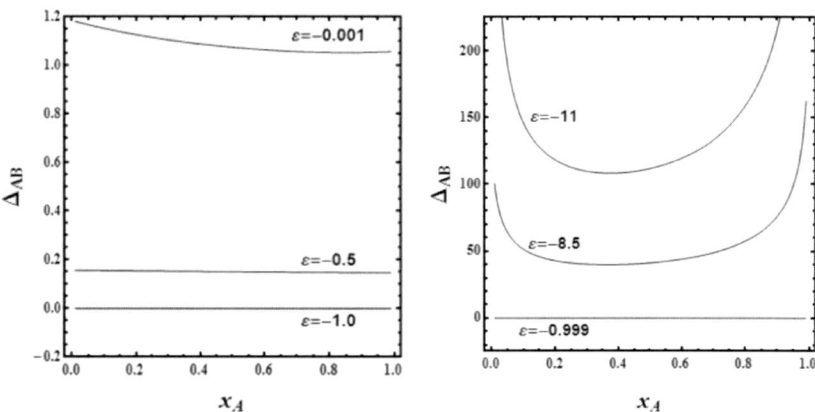

Fig. 9.18 The measure of the deviation of *SI* solution Δ_{AB} for the same system as in Fig. 9.10.

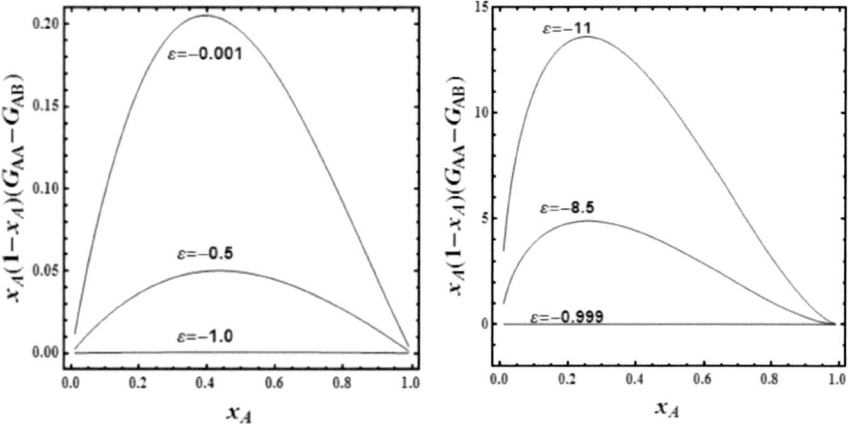

Fig. 9.19 The limiting coefficient of the preferential solvation of A with respect to A and B for the same system as in Fig. 9.10.

These quantities are of interest in their own right. First, we note that for $|\varepsilon| = 1$ values of δ_{AA}^0 are zeros and independent of x_A. This result is due to the fact that at $|\varepsilon| = 1$, the two components are identical, and therefore there is no

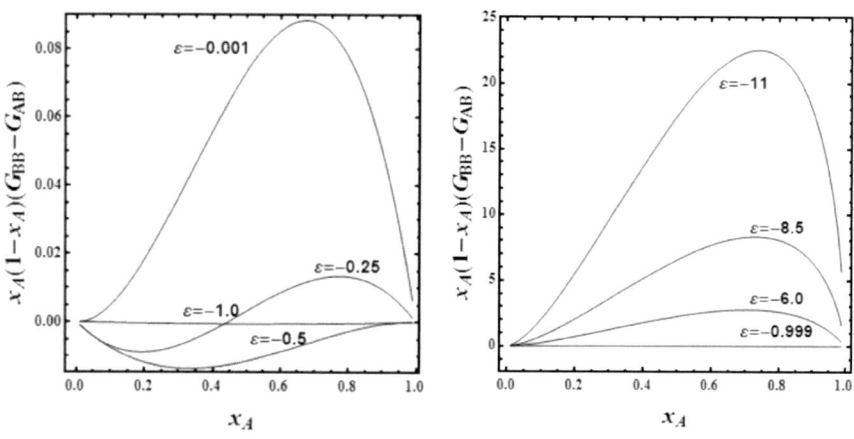

Fig. 9.20 The limiting coefficient of the preferential solvation of B with respect to A and B for the same system as in Fig. 9.10.

preferential solvation. It should be noted that this lack of PS is not a result of the SI behavior but a result of the *identity* of the particles. In general, SI solutions can occur even for different particles, but SI behavior does not imply zero PS. The limiting coefficient of the PS of B with respect to A and B is shown in Fig. 9.20. Two interesting features that distinguish these curves from the ones in Fig. 9.19 are: First, we find both positive and negative values of the PS; second, the sign of the PS changes when the composition of the mixture changes. This phenomenon is found in some aqueous solutions.[2]

Finally, we present some values of the solvation Gibbs energies of A and B in the entire range of compositions.

Figure 9.21 shows ΔG_A^* as a function of composition. Note again that for $|\varepsilon| = 1$, ΔG_A^* is independent of the composition. This is a result of the identity of the particles. Note also that the value of ΔG_A^* is non-zero even at $|\varepsilon| = 1$.

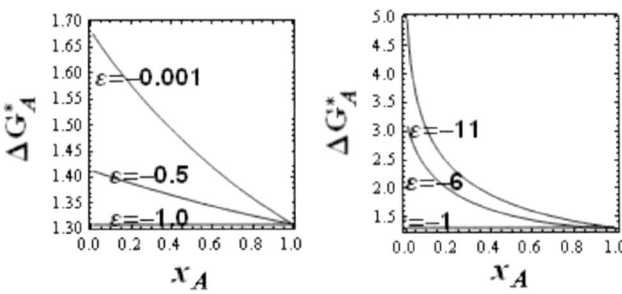

Fig. 9.21 The solvation Gibbs energy ΔG_A^* in the same system as in Fig. 9.10.

ΔG_A^* measures the average value of the quantity $\exp[-\beta B_A]$, where B_A is the total interaction energy of a single A with the rest of the system. This quantity is often referred to as the "free energy of interaction" of the species A. As $|\varepsilon| \leq 1$ becomes smaller, we see that the values of ΔG_A^* become larger. On the other hand, for $|\varepsilon| \geq 1$, the reverse trend is true. Note also that in all cases, ΔG_A^* converge to the value of ΔG_A^* for the solvation of A in pure A (i.e. at $x_A \to 1$). We note also that the behavior of ΔG_A^* as a function of x_A for $|\varepsilon| \geq 1$ is similar to the case argon in mixtures of argon and xenon.[2] The interpretation of this phenomena is simple. Starting from $x_A = 0$, we have the solvation of A surrounded by pure B. As x_A increases from $x_A = 0$ to $x_A = 1$, the surroundings of A change from all B to all A particles, hence, the free energy of interaction decreases strongly towards the value of ΔG_A^* for pure A. This interpretation also applies to the solvation Gibbs energies of argon in mixtures of argon and xenon.

A different behavior is exhibited by ΔG_B^* shown in Fig. 9.22. Here again, we find that ΔG_B^* is composition

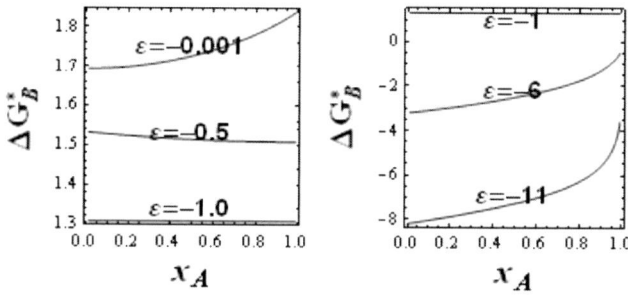

Fig. 9.22 The solvation Gibbs energy ΔG_B^* in the same system as in Fig. 9.10.

independent for $|\varepsilon| = 1$. However, in contrast to the case of ΔG_A^* (Figure 4.24), the values of ΔG_B^* do not converge to a single value for pure $B(x_A = 0)$. The reason is simple. In the case of ΔG_A^*, *pure A* (i.e. $x_A = 1$) means a unique system of A particles with fixed molecular parameters $\sigma_{AA} = 1$ and $\varepsilon_{AA} = 1$. On the other hand, when we say pure B (i.e. $x_A = 0$), there are different "pure" systems with $\sigma_{BB} = 1$, but varying values of ε_{BB}.

The systems studied in this section are quite simple. Nevertheless, the main message as expressed in the introductory section is quite clear. The *local* properties of the mixtures are richer and more informative than the *global* properties. This conclusion becomes *a fortiori* true for more complex mixtures, particularly aqueous solutions which we shall study in the next chapter.

10

A Study of Water and Aqueous Solutions in 1D

Water is the most familiar liquid. It is also the most unusual liquid, and of course it is the medium in which life has emerged and evolved. It therefore does not come as a surprise that liquid water has been, and continues to be, the most studied liquid.

In this chapter we will focus on a few aspects of liquid water where the power of statistical thermodynamics is demonstrated. Before doing so, it should be said that nowadays most studies on liquid water and aqueous solutions are carried out on computers. We will not discuss any of these studies here. Instead, we chose a few questions for which statistical mechanics provides exact answers. This is more than enough to convince the reader of the power of statistical mechanics in answering difficult questions relevant to real and vital systems.

About forty years ago when statistical mechanical methods were first applied to the study of liquid water, it was

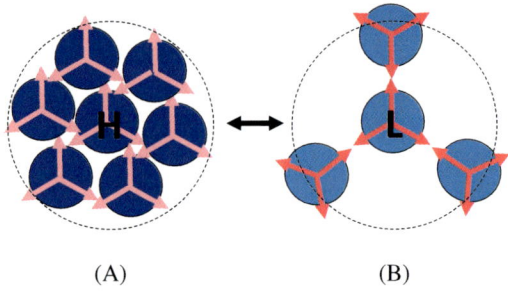

(A)   (B)

Fig. 10.1 A schematic description of the two possible modes of packing of water molecules. (A) A closed packed configuration, and (B) An open-structure of hydrogen-bonded molecules. The fluctuation between these two kinds of configuration is referred to as the "principle".

conjectured that there exists a general principle underlying some of the most unusual properties of water and its solutions. This principle states that if you take a snapshot of liquid water you will observe water molecules in different environments characterized by their local densities. Two extreme cases are shown in Fig. 10.1. In Fig. 10.1A, you see that if you are situated on the central molecule denoted H, you will see many molecules around you. On the other hand, if you were sitting on a molecule denoted L in Fig. 10.1B, you will see fewer molecules around you.

This phenomenon is of course not specific to water. Take a snapshot of the configuration of any liquid, and you will find molecules which have *high* and *low* local density. This is shown in Fig. 10.2.

What is unusual in water is that molecules which are in *low local* density have on average *stronger binding energies* to their surrounding molecules than molecules which are in *high local* density. In any normal liquid, at not too high densities

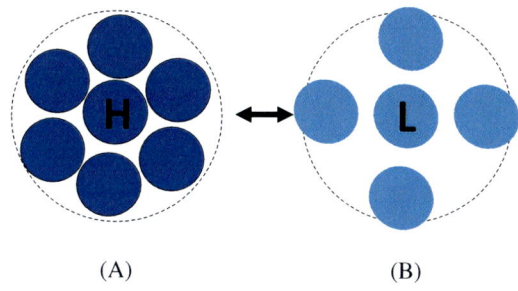

Fig. 10.2 The two possible modes of packing in a "normal" liquid. (A) Closed packed, (B) Open structure.

we should find the "normal" correlation between *low local* density and weaker binding interactions.

This "normal" correlation is easy to understand. The more molecules are in the neighborhood, the more interaction between the central molecule and its environment. This is shown schematically in Fig. 10.2. In water, on the other hand, molecules in a high local density have also more neighbors. However, the interaction energy between these molecules is not as high as a molecule which forms hydrogen-bond (HB). Formation of maximum hydrogen bonds, hence, also maximum binding energy, can be achieved in a relatively low local density.

Fifty years ago it was believed that the peculiar properties of water were due to the characteristic tetrahedral structure around each water molecule in pure ice. However, later it was found that the specific correlation between *local* density and binding energy was more important, and more fundamental than the tetrahedral geometry in liquid water.

This idea was first tested in a 2D model of water-like particles. What is most remarkable is that this correlation between local density and binding energy was also responsible for the most outstanding properties of water and aqueous solutions. For this reason it was found fitting to refer to this unusual correlation as the *principle* of *water*.[1]

Originally, it was only suspected that this principle was related to the properties of water. It was only twenty years later that this conjecture was scrutinized and put under the test of statistical mechanics. Here is the formulation of the question. If the *principle* of correlation between low local density and high binding energy is responsible for the outstanding properties of water, then we must be able to test it in any system in which this principle can be implemented. Obviously, the simplest system in which the principle can be implemented is a 1D liquid. In a 1D system a model can be constructed in which the *principle* is a built-in feature, and for which all the thermodynamics of the system can be calculated exactly. Hence, the conjecture can be proven. In the next section, we shall describe a 1D model and its solution. We will examine only one property of liquid water, the unusual temperature dependence of the volume of water between $0-4°C$. In Section 10.3 we will examine one unusual property of aqueous solutions; the large negative entropy of solvation of inert-molecules in water. As we will see, both these seemingly unrelated phenomena are in fact related through the *principle* of water.

10.1 A Primitive Model for Water-Like Particles

During the early 1970s, a few models for a water molecule were suggested. It was believed that the *structure* of water is an essential aspect of the properties of water, as a result of which, all models of liquid water were designed with the aim of obtaining the characteristic tetrahedral structure around a water molecule. The first attempt to challenge the essentiality of the tetrahedral structure was to construct a 2D model of water-like particles.[1] In this 2D model the structure, not necessarily the tetrahedral-structure was still an important aspect of the model. However, it was soon discovered that the structure per-se was not essential for understanding the properties of liquid water. Instead, the *principle* of *water* was deemed to be a more essential aspect. Unfortunately, this could not be tested in the 2D model. Therefore, a 1D model of water was constructed where the *structure* does not play any role.

A one-dimensional model for water may sound as an extremely unrealistic model for such a complex liquid. Indeed, it is! If the aim is to calculate thermodynamic quantities and to compare them to the corresponding experimental quantities of real water, then clearly, a 1D model is useless. However, if the aim is to *understand* the outstanding properties of water on a molecular level, then any model in any dimension could be useful.

Liquid water is known to exhibit some outstanding properties, often referred to as anomalous properties, such as the negative temperature dependence of the volume, the high heat capacity, the minimum in the isothermal compressibility and many more.

It was long recognized that these outstanding properties of water are related to the *open-structure* or the *ice-like* character of liquid water. The idea that water contains some kind of *structure*, related to the structure of ice, is very old. It first appeared explicitly in writing by Roentgen in 1892.[2]

In all these models, the *structural* aspect of liquid water was traditionally emphasized as the main molecular *reason* for the anomalous behavior exhibited by liquid water. However, underlying this, relatively ill-defined concept of structure (which was much later defined in statistical mechanical terms), lies a more fundamental principle, which can be defined in molecular terms, and which does not explicitly mention the concept of structure, yet this principle is responsible for the unusual properties of liquid water. This principle states that there is a range of temperatures and pressures at which the water–water interactions produce a unique correlation between high local density, and a weak binding energy. Clearly, this principle does not depend on the concept of *structure*. As will be demonstrated in this chapter, it is this principle, not the *structure* per se, which is responsible for the unique properties of water as well as of aqueous solutions.

In this chapter, we present two simplified models that manifest many of the outstanding properties of water and aqueous solutions. The first is referred to as the primitive 1D model, a simplified version of the earlier 1D model published in 1992.[2] It has the advantage of bearing nothing but the "principle" mentioned above. It is shown that a double-well feature of the pair potential is not really necessary. The implementation of the principle is achieved by defining only two distances; the hard core "diameter", σ_1 and the "hydrogen-bonding" (HB), distance σ_2, with the requirement that $\sigma_2 > \sigma_1$.

The second model is a simplified version of a model published in 1969.[3] It is referred to as the primitive cluster

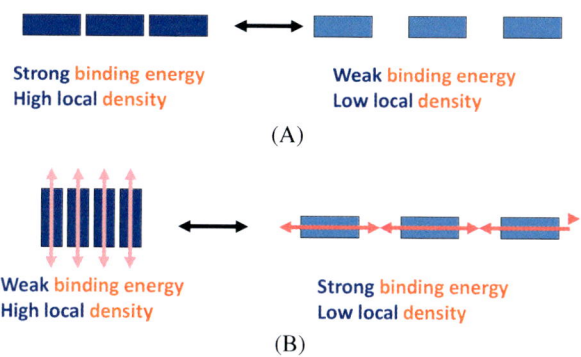

Fig. 10.3 A schematic description of the 1D primitive model. (A) Normal liquid, (B) Water-like particles.

model for water. This model produces the same results for water, as the primitive model. However, it has two advantages. First, one can follow the "molecular mechanism" underlying the characteristic behavior of liquid water. Second, it is easier to apply it in the study of aqueous solutions of simple solutes.

10.1.1 (***) The Primitive Model and Its Partition Function

The system consists of N water-like particles, denoted W, in one dimension (Fig. 10.3). We consider only nearest neighbors interactions between pairs of consecutive particles. The potential function is described in Fig. 10.4C, and is defined as

$$U_{HB}(R) = \begin{cases} \infty & 0 \leq R < \sigma_1 \\ 0 & \sigma_1 \leq R < \sigma_2 \\ \varepsilon_{HB} & \sigma_2 \leq R < \sigma_2 + \delta_2 \\ 0 & R \geq \sigma_2 + \delta_2 \end{cases} \tag{10.1}$$

Here, σ_1 is the hard-rod "diameter" of the particles. Between σ_1 and $\sigma_2 > \sigma_1$, there is a flat region (in previous

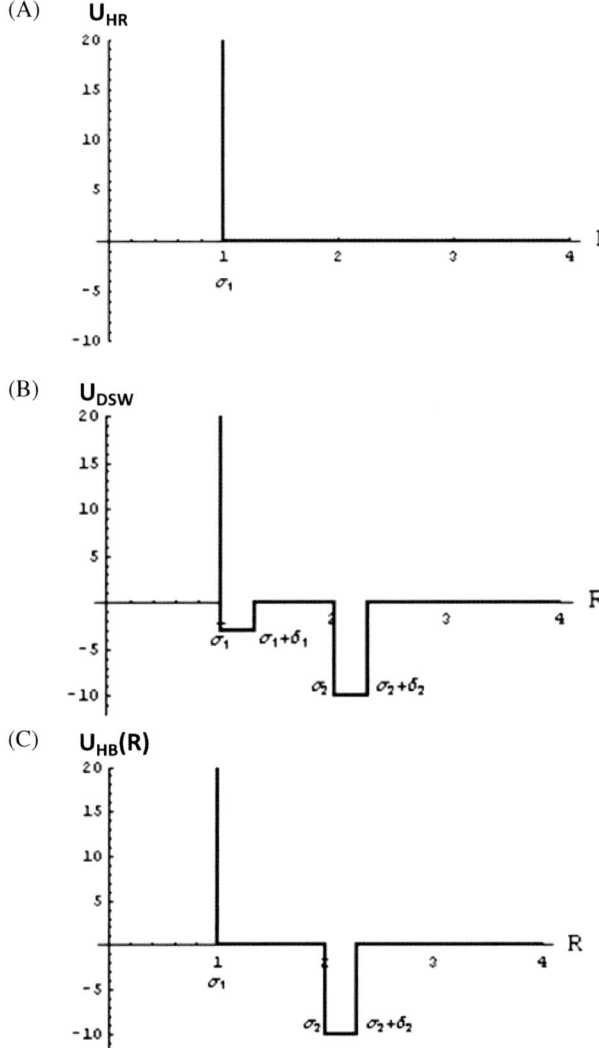

Fig. 10.4 The various pair potentials used in the study of water in 1D. (A) The hard rod (HR) potential, (B) The double square well (DSW) potential used earlier in the study of water. (C) The pair potential for the primitive model, referred to as the hydrogen bond(HB) potential.

models this region contained an additional square well potential, see Fig. 10.4). ε_{HB} is referred to as the hydrogen-bond (HB) energy and its range is between σ_2 and $\sigma_2 + \delta_2$. In this chapter, we use dimensionless parameters. The Planck constant and the Boltzmann constant are chosen as unity, hence ε_{HB} has the same units as the temperature T.

It is already clear from the Eq. (10.1) and from Fig. 10.4, that this potential fulfills the requirement of the "principle of water". Particles that are closer to each other (in the range σ_1 to σ_2) experience weaker binding energy, while at larger distances (in the range σ_2 to $\sigma_2 + \delta_2$), experience a stronger binding energy. This is enough to produce the required correlation between low *local density* and strong *binding energy*.

We describe here briefly the construction of the partition function (PF) of this system[1]

The total interaction energy is written as

$$U_N(R_1, R_2, \ldots, R_N) = \sum_{i=1}^{N-1} U_{HB}(R_{i,i+1}) \qquad (10.2)$$

where R_i is the location of the ith particles, and $R_{i,i+1}$ is the distance between two consecutive particles i and $i + 1$. As before we "close" the cycle by identifying the Nth particle with the first one.

The classical canonical PF is

$$Q(T, L, N) = \frac{1}{\Lambda^N N!} \int_0^L \cdots \int_0^L dR^N$$

$$\times \exp[-\beta U_N(R^N)] \qquad (10.3)$$

where $\beta = T^{-1}$, Λ is the momentum PF in one dimension, which in our case, is

$$\Lambda = \frac{h}{\sqrt{2\pi m k_B T}} \qquad (10.4)$$

(In the primitive model, we choose $h = 1$, $k_B = 1$ and $m = 1$). The integration for each particle ranges from 0 to L. As we have done in Chapter 8, we impose an order on the particles, and the configurational PF can be rewritten as

$$\frac{1}{N!} \int_0^L \cdots \int_0^L d\mathbf{R}^N \exp[-\beta U_N(\mathbf{R}^N)]$$

$$= \int_0^L d\mathbf{R}^N \int_0^{R_N} dR_{N-1} \cdots \int_0^{R_2} dR_1$$

$$\times \exp[-\beta U_N(\mathbf{R}^N)] \qquad (10.5)$$

Next, we take the Laplace transform of the rhs of (10.5), which is essentially the same as transforming from the variables T, L, N to the variables T, P, N where P is the one-dimensional pressure. Denote by

$$\psi_{WW}(P, T) = \int_0^\infty dR \exp[-\beta P R - \beta U_{HB}(R)] \qquad (10.6)$$

We can apply the convolution theorem to obtain the isothermal-isobaric PF:

$$\Delta(T, P, N) = \psi_{WW}(P, T)^N / \Lambda^N \qquad (10.7)$$

Thus, for any given pair potential $U(R)$, we can calculate the PF in (10.7), and therefore all the thermodynamic quantities from the fundamental relation between $\Delta(T, P, N)$ and the Gibbs energy of the system

$$G(T, P, N) = -T \ln \Delta(T, P, N) \qquad (10.8)$$

10.1.2 Selected Illustrative Results for the Primitive Model

In this section, we present some results for the thermodynamic quantities of the 1D model of particles interacting through the potential function (10.1). We will compare these results with results for a reference liquid, which consists of hard-rods (HR). The interaction between the particles in the "normal", reference system, is Fig. 10.4.

$$U_{HR}(R) = \begin{cases} \infty & 0 \le R < \sigma_1 \\ 0 & R > \sigma_1 \end{cases} \qquad (10.9)$$

Hence

$$\psi_{HR} = \int_0^\infty dR \exp[-\beta PR - \beta U_{HR}(R)] = \frac{\exp[-\beta P \sigma_1]}{\beta P} \qquad (10.10)$$

We will refer to the reference system of the hard-rod system (HR) and to the water-like system as the hydrogen-bonded (HB) system.

The most unusual, and well-known property of liquid water is the temperature dependence of the molar volume between 0 to 4°C.

The molar "volume" is given by

$$\langle l \rangle = \frac{\langle L \rangle}{N} = \frac{1}{N} \frac{\partial G}{\partial P} = \frac{\partial \mu}{\partial P} \qquad (10.11)$$

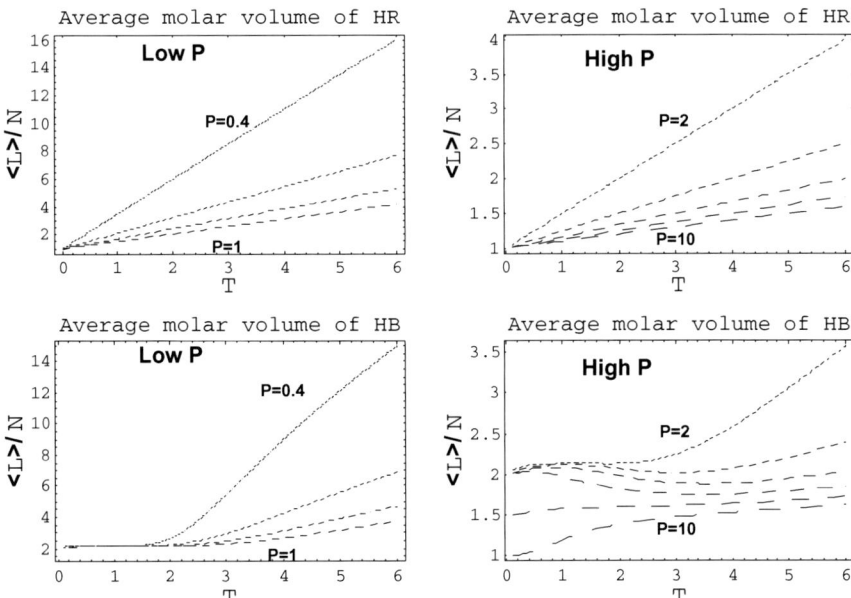

Fig. 10.5 The temperature dependence of the average length per particle (the molar "volume") as a function of the temperature for a few values of the pressure. Upper panel correspond to the reference hard particles or hard rods (HR), and the lower panel to the hydrogen bonded (HB) water-like system. In this figure, as well as in all the following figures we indicate only a few values of the pressure next to the corresponding curves. The in-between curves have in-between pressures.

In Fig. 10.5, we show the temperature dependence of the average length per particle $\langle L \rangle / N$ as a function of temperature for the following parameters.

$$\sigma_1 = 1, \quad \sigma_2 = 2, \quad \delta_2 = 0.3 \quad \varepsilon_{HB} = -10 \quad (10.12)$$

As expected, for the hard rod system, the average volume always increases with temperature both at low and at high pressures. Note also that in the limit of $T \to 0$, the molar volume of the HR system tends to $\sigma_1 = 1$, as it should. On the other hand, for the water-like molecules interacting through the pair potential $U_{HB}(R)$, Eq. (10.1) we find quite a different behavior.

First, note that at low pressure and high temperature, the system behaves "normally", i.e. the molar volume increases monotonically with increasing temperature, or decreasing pressure. On the other hand, in the limit of very low temperatures $T \to 0$, we have two distinctly different limiting volumes. At low pressures, the limit of the molar volume tends to σ_2 (here $\sigma_2 = 2$). This corresponds to the "ice-like" open structure, where all pairs attain the distance between σ_2 and $\sigma_2 + \delta_2$. On the other hand, at higher pressure (here for about $P \approx 9$), the limiting value of the molar volume is that of the closed packed system as in the case of hard rods system.

The most interesting behavior. which is a direct result of the choice of the particular pair potential $U_{HB}(R)$ in (10.1) is for pressures in the range $4 \le P \le 9$, and temperatures in the range of $2 \le T \le 5$. In this range we observe a minimum in the volume as a function of temperature. It is quite surprising that such a "primitive" model exhibits this outstanding behavior.

Note that as the pressure increases, the location of the minimum initially shifts to the right (i.e. to higher temperature), then to the left but becomes more and more shallow. At much higher pressures ($P \ge 9$), the minimum disappears, and the system's volume behaves "normally" (as is the case for real liquid water at high pressures).

More details of the molar volume as a function of pressure near the minimum are shown in Fig. 10.6.

The pressure dependence of the molar volume of HR, and the water-like system is shown in Fig. 10.7. Note, that

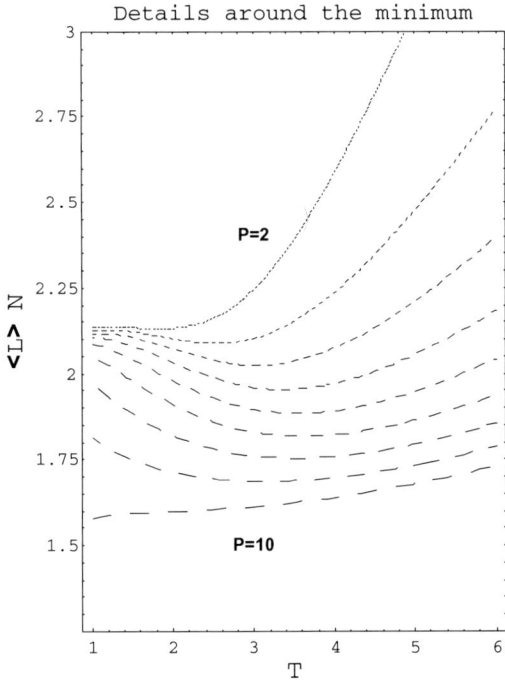

Fig. 10.6 Details of the temperature dependence of the molar volume near the minimum of the volume.

for the *HR* system, molar volume is always a monotonically decreasing function of pressure. On the other hand, for the water-like system, as we decrease the temperature, molar volume changes sharply (similar to a phase-transition) from the open structure (of molar volume $\sigma_2 = 2$) to the close-packed structure (of molar volume $\sigma_1 = 1$). As the temperature is lowered, the transition becomes sharper and sharper.

Figure 10.8 shows the "isotope" effect on the temperature of minimum volume of the system for three different values of the *HB* energy; $\varepsilon_{HB} = -8, -9, -10$. The results shown in Fig. 10.8 are qualitatively similar to the corresponding experimental data on H_2O, D_2O and T_2O. First, upon

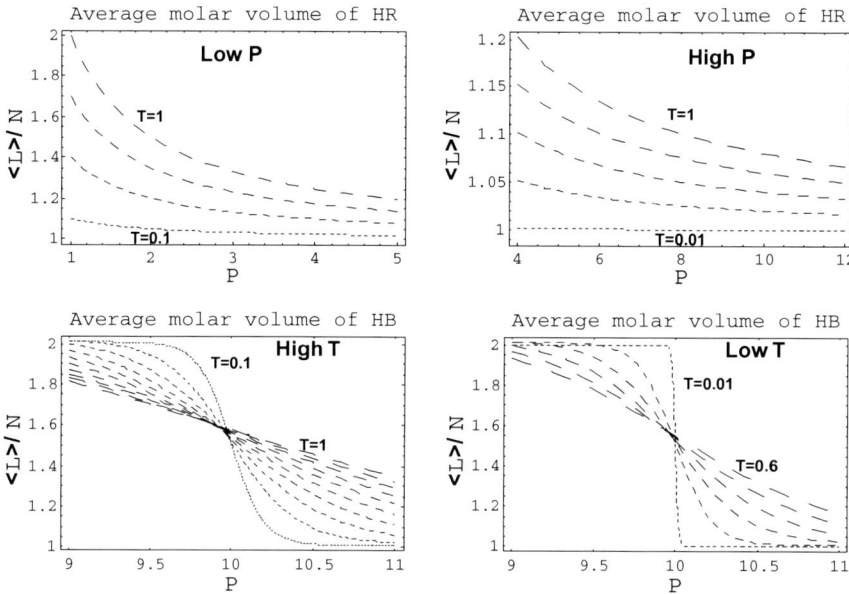

Fig. 10.7 The pressure dependence of the molar volume for different values of the temperature.

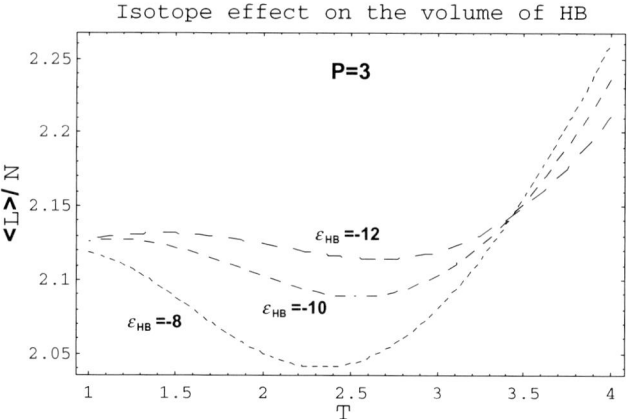

Fig. 10.8 The "isotope" effect on the molar volume near the minimum, for three values of the *HB* energy ε_{HB}.

increasing the *HB* energy (keeping all other parameters constant), the location of the minimum is shifted to the right (It is well-known that H_2O and D_2O have almost exactly the same geometries, i.e. the *OH* distance and the *HOH* angles are the same. The only significant difference is the strength of the *HB* energy, which in the 1D model is represented by the parameter ε_{HB}). Second, the molar volume of the three "isotopes", in the neighborhood of the minimum, increases with the strength of the *HB* energy. This reflects the higher degree of "structure" of the system having the stronger *HB*'s. This feature is similar to the isotope effect in real water.

Finally, Fig. 10.9 shows the "isotope" effect on the pressure dependence of the volume near the phase transition. We choose one low temperature, $T = 0.01$, and plot the volume as a function of pressure for the three values of $\varepsilon_{HB} = -8$, $-10, -12$. In the three cases, there is a sharp transition from

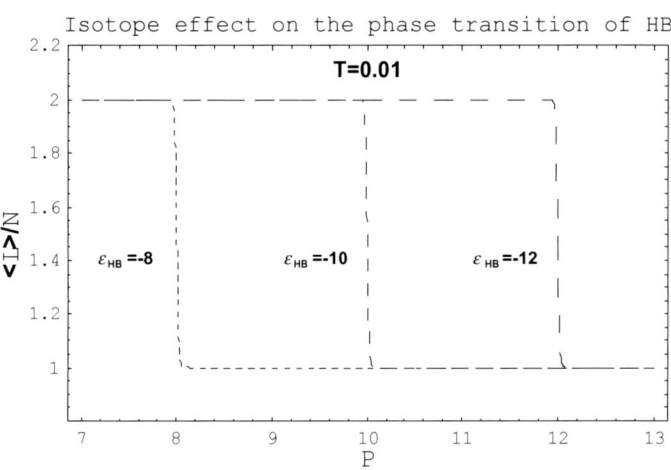

Fig. 10.9 The "isotope" effect on the behavior of the molar volume near the phase transition.

the open-structure to the close-packed structure as we increase the pressure. The location of the transition pressure is different for the different case. As $|\varepsilon_{HB}|$ increases, i.e. stronger *HB*, the "open-structure" becomes more stable, and a higher pressure is required to produce the transition.

In conclusion we can say that implementation of the "principle of water" in the most primitive model leads to one of the most outstanding properties of liquid water; the negative temperature dependence of volume of water between 0 and 4°C. The emphasis in this section was not on the properties of liquid water, but in demonstrating the power of statistical thermodynamics in verifying a conjecture, in this case the relationship between the "principle of water," and the unusual properties of water. More details on the properties of water and attempts at interpreting these properties may be found elsewhere.[1] In the next section, we shall discuss another outstanding property of water, the unusually large and negative entropy of solvation of inert molecules in water.

10.1.3 (***) *The Primitive Cluster Model for Water*

In this section, we introduce a model which is essentially equivalent to the one described in Section 10.1.1. This model, referred to as the primitive cluster model has several features that makes it more useful in the study of the molecular reasons underlying the anomalous behavior of liquid water and aqueous solutions. For pure water, as we shall see below, the two models give essentially the same results. However, with the cluster model, we can get a deeper insight into the "mechanism" underlying the anomalies of water, namely, the "structural changes" (here, essentially the change in the cluster-size distribution) in the liquid that leads to the anomalous behavior. As we will see in Section 10.2, the cluster-model is also more convenient for the study of some of the most outstanding properties of aqueous solutions of simple solutes.

This is a simplified version of a model first introduced by Lovett and Ben-Naim (1969). The idea is to define a sequence of n hydrogen-bonded particle as an n-cluster (mimicking the ice-like clusters of HB'ed molecules in liquid water). In doing so, we include all the HB's in the cluster as part of the "internal" partition function of the n-cluster. The interaction potential between any pair of clusters (including the 1-cluster, i.e., the monomers, or non HB ed particle) is now the hard rod (HR) potential. In the original model, this part was chosen to be a square-well potential. As with the primitive model discussed in Section 10.1.1, the primitive version of the model consists of removing the square-well potential and leaving only the hard-core repulsive part.

We present a brief derivation of the partition function for the cluster model. It is basically the same as the one derived by Lovett and Ben-Naim, except for the simplification of the pair interactions. In the rest of this section we present some technical details about the solution of the PF of the system. The reader who is not interested in these details can skip this part.

The system consists of monomers, the interaction potential between them is the hard-rod potential as described in Section 10.1.1. In addition, we have clusters of different sizes. An n-cluster consists of n monomers connected by "hydrogen-bonds". Once we have defined a cluster as a sequence of *HB*'ed monomers, the system may be viewed as a mixture of different species; monomers, 2-clusters, 3-clusters, etc. The grand partition function for such a fluid mixture is

$$\Xi(T,L,\lambda_1,\ldots\lambda_n)$$

$$= \sum_{N_1,N_2,\ldots N_n=0}^{\infty} \prod_{i=1}^{n} \frac{\lambda_i^{N_i}}{N_1!\Lambda_i^{N_i}} \int_0^L \cdots \int_0^L dR^{N_1} \cdots dR^{N_n}$$

$$\times \exp[-\beta U_N] \tag{10.13}$$

where U_N is the total potential energy of interaction in the system.

For the *HB*-potential, we simply take a harmonic potential of the form $U_{HB}(R) = \frac{1}{2}k(R-a)^2$, and an energy parameter ε_{HB}. We also denote by

$$\eta(T,P) = \exp[-\beta\varepsilon_{HB}]\psi_{HB}$$

$$= \exp[-\beta\varepsilon_{HB}] \int_0^{\infty} dR \exp[-\beta PR - \beta\frac{k}{2}(R-a)^2] \tag{10.14}$$

There is a standard procedure to proceed from (10.13) to obtain the Laplace transform of the grand partition function with respect to the length variable L, i.e.

$$\psi[\Xi(T,L,\lambda)] = \int_0^{\infty} \exp[-\beta PL]\Xi(T,L,\lambda) \tag{10.15}$$

By choosing a specific ordering of the species, then applying the convolution theorem, the Laplace transform of the grand partition function can be rewritten

$$\psi[\Xi(T,L,\lambda)] = \sum_{N=0}^{\infty} Tr[M^N] \tag{10.16}$$

where $Tr[M^N]$ is the trace of the matrix M^N. The elements of the matrix M are defined by

$$M_{1,1} = \frac{\lambda_1}{\Lambda_1}\psi_{HR}$$

$$M_{1,n} = \left[\frac{\lambda_1\lambda_n}{\Lambda_1^{n+1}}\psi_{HB}^{n-1}\exp[-\beta(n-1)\varepsilon_{HB}]\right]^{1/2}\psi_{HR}$$

$$M_{n,m} = \left[\frac{\lambda_n\lambda_m}{\Lambda_1^{n+m}}\psi_{HB}^{n+m-1}\exp[-\beta(n+m-1)\varepsilon_{HB}]\right]^{1/2}\psi_{HR} \tag{10.17}$$

where ψ_{HB} is defined in (10.14).

If γ_i is the ith eigenvalue of the matrix M, then we can rewrite (10.16) as

$$\psi[\Xi(T,L,\lambda)] = \sum_{i}\sum_{N=0}^{\infty}\gamma_i^N = \sum_{i}(1-\gamma_i)^{-1} \tag{10.18}$$

Normally, to recover the grand partition function, one needs to take the inverse Laplace transform of (10.18). However, in our case, we can view the Laplace transform of $\Xi(T,L,\lambda)$ as the "generalized" partition function, and we can obtain all the thermodynamic quantities of the system from the eigenvalues of the matrix M.

Define the auxiliary quantities

$$t_1^2 = \frac{\lambda_1}{\Lambda_1} \quad \text{and} \quad t_n^2 = \frac{\lambda_n}{\Lambda^n} \psi_{HB}^{n-1} \exp[-\beta(n-1)\varepsilon_{HB}]$$

(10.19)

The matrix M can be transformed into

$$M = \psi_{HR} \begin{pmatrix} t_1^2 & t_1 t_2 & t_1 t_3 \\ t_1 t_2 & t_1^2 & t_1 t_3 \\ \vdots & \vdots & \vdots \end{pmatrix}$$

(10.20)

Fortunately, this matrix can be transformed by an orthogonal transformation into a matrix of the form

$$M' = \psi_{HR} \begin{pmatrix} \sum t_1^2 & 0 & 0 & 0 \\ 0 & 0 & 0 & 0 \\ 0 & 0 & 0 & 0 \end{pmatrix}$$

(10.21)

Clearly, this matrix has only one eigenvalue, γ, which is obtained from the equation

$$\psi_{HR} t^2 - \gamma = 0$$

(10.22)

where $t^2 = \sum_{i=1}^{\infty} t_i^2$.

In order to use the Laplace transform of the grand partition function in (10.18), γ must be equal to 1, hence, the sum on the rhs of (10.18) diverges. This leads to the implicit equation of state of the form

$$\psi_{HR}(T, P) t^2(T, P, \lambda) = 1$$

(10.23)

The condition of chemical equilibrium among all the species is $\mu_n = i\mu_1$, or equivalently $\lambda_n = \lambda_1^i$, where λ_1 is the absolute activity of the monomers (which is also the absolute activity of water when all the clusters are at equilibrium). After inserting the condition of chemical equilibrium into t^2, we can carry out the summation over all species i, to obtain

$$t^2 = \sum_{i=1}^{\infty} t_i^2 = \frac{\lambda_1}{\Lambda_1} + \sum_{i=2}^{\infty} \frac{\lambda_1^n}{\Lambda_1^n} \eta^{n-1}$$

$$= \frac{\lambda_1/\Lambda_1}{1 - \eta\lambda_1/\Lambda_1} \qquad (10.24)$$

At this stage we have only one absolute activity, we write $\lambda = \lambda_1$ and rewrite the equation of state (10.23) as

$$\psi_{HR}(T,P)\frac{\lambda/\Lambda_1(T)}{1 - \eta(T,P)\lambda/\Lambda_1(T)} = 1 \qquad (10.25)$$

This is an implicit equation from which we can eliminate the chemical potential μ in the form

$$\mu(T,P) = T \ln \lambda = T \ln \left[\frac{\Lambda_1}{\psi_{HR} + \eta} \right] \qquad (10.26)$$

where

$$\psi_{HR} = \int_0^{\infty} \exp[-\beta PR] \exp[-\beta U_{HR}(R)] = \frac{\exp[-\beta P\sigma_1]}{\beta P} \qquad (10.27)$$

From the chemical potential in (10.26) we can get all the required thermodynamic quantities of the system. As we have pointed out earlier, by choosing σ_1 and ε_{HB} as in the primitive model and choosing a to be equal to $\sigma_2 + \delta_2/2$, in the primitive model, there remains only one parameter, the force constant k to be chosen in such a way that the results of this model will be almost identical with the results obtained in Section 10.1.2.

Formally, the identity of the two models, for a fixed T and P, can be obtained by requiring that ψ_{WW} of Section 10.1.1 be equal to

$$\psi_{WW} = \psi_{HR} + \eta \qquad (10.28)$$

Once we have chosen the parameters σ_1, $\sigma_2 = a$, δ_2 and ε_{HB}, and with some experimentation, one can get almost identical results as the ones obtained in Section 10.1.2. Therefore, we shall not describe the thermodynamics of this system. Here, we present some new results on the cluster size distribution of this model. We will revert to this model in Section 10.2 to study aqueous solutions.

10.1.4 Cluster's Size Distribution in the Primitive Cluster Model

We now turn to examine the cluster's size distribution and its dependence on T and P. To do this, we need to go back to the equation of states in (10.23) *before* the insertion of the equilibrium condition. We write the equation of states as

$$F(T,P,\lambda) = \psi_{HR}(T,P) \sum_{i=1}^{\infty} t_i^2(T,P,\lambda_i) = 1 \qquad (10.29)$$

where all the λ_i are considered as independent variables (this is equivalent to a system where the equilibrium among all the species is "frozen in"). The Gibbs–Duhem relation is

$$S\,dT - L\,dP + \sum N_i\,d\mu_i = 0 \qquad (10.30)$$

from which we can obtain all the required densities

$$\rho_i = \frac{N_i}{\langle L \rangle} = -\frac{\partial F / \partial \mu_i}{\partial F / \partial P} = \frac{-\frac{\partial F}{\partial \lambda_i} \frac{\partial \lambda_i}{\partial \mu_i}}{\partial F / \partial P} \tag{10.31}$$

After the differentiation of F with respect to λ_i, we can insert the condition of equilibrium $\lambda_i = \lambda^i$ to obtain the densities of all the species at equilibrium, this leads to

$$\rho_j = \frac{-\beta \lambda^j \eta(T, P)^{j-1}}{\Lambda^j \partial F / \partial P} \tag{10.32}$$

The total number of water molecules is

$$\rho_W = \sum_{i=1}^{\infty} i \rho_i \tag{10.33}$$

We define the mole fraction of the j-clusters as

$$X(j) = \frac{\rho_j}{\sum_{i=1}^{\infty} i \rho_i} \tag{10.34}$$

Note that the normalization of $X(j)$ is

$$\sum_{i=1}^{\infty} j X(j) = 1 \tag{10.35}$$

Since an n-cluster contains $(n - 1)$ hydrogen bonds, the average number of HB per particle is

$$\langle HB \rangle = \frac{\sum (j - 1) \rho_j}{\sum i \rho_i} = \sum (j - 1) X(j) \tag{10.36}$$

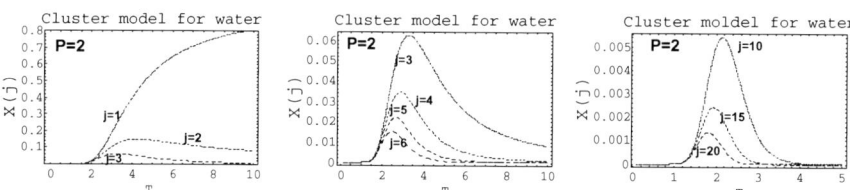

Fig. 10.10 Dependence of $X(j)$ on the temperature, for some values of j.

It is easily shown that $X(j)$ is related to $\langle HB \rangle$ by the equation, see Ben-Naim (2009)

$$X(j) = \frac{(1 - \langle HB \rangle)^2}{\langle HB \rangle^{j-1}} \tag{10.37}$$

In the following, we present some results for the cluster-size distribution and the related quantities. The calculations are performed for the following set of parameters

$$\sigma_1 = 1, \quad \sigma_2 = a = 2, \quad k = 1000, \quad \varepsilon_{HB} = -10 \tag{10.38}$$

Figure 10.10 shows the temperature dependence of the concentration of each type of cluster. As expected, the mole fraction of the monomers increases monotonically as a function of T, eventually reaching $X(1) = 1$ at high temperatures.

For $j \neq 1$, we have initially an increase in the mole fraction $X(j)$, but eventually decreases at very high temperature. As expected, the larger the cluster size, the lower the maximum of $X(j)$. Figure 10.10 shows the temperature dependence of $X(j)$ for three sets of cluster sizes: small, intermediate and large.

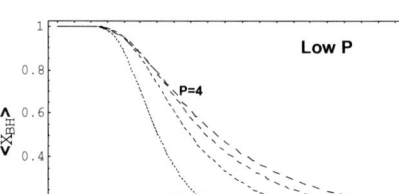 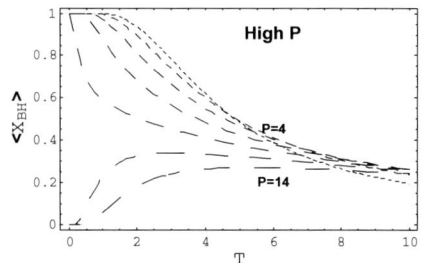

Fig. 10.11 The average number of hydrogen bonds per molecule $\langle HB \rangle$ as a function of T at different pressures.

The overall "structural changes" in this system are captured by the quantity $\langle HB \rangle$ defined in (10.36). This is the analogue to the average number of HB per particle, in real liquid water. This quantity may be used as a measure of the overall *structure* of the system.

Figure 10.11 shows the temperature dependence of $\langle HB \rangle$ for several values of the pressure. Note that at low temperatures ($T \leq 2$), we start with $\langle HB \rangle \approx 1$, i.e. almost all the particles in the system are HB'ed. Increasing the pressure causes a breakdown of HBs, hence decreasing $\langle HB \rangle$. At higher temperature ($T \geq 3$), this trend is reversed, i.e. increasing the pressure causes formation of more HBs. This behavior is typical for low pressures $P \leq 10$. At higher pressures, ($P \geq 11$), and low temperatures, all the particles are close-packed with no HB's. Hence, increasing the temperature causes an increase in $\langle HB \rangle$.

In order to gain insight into the molecular mechanism underlying the anomalous behavior of liquid water (as well as of aqueous solutions, see Section 10.2), we need to examine the temperature and the pressure dependence of the quantity $\langle HB \rangle$. To see that, it is sufficient to consider a simple two-structure model for water. Let x_{HB} be the mole fraction of the *HB*'ed component (or the ice-like species in the traditional mixture model theories of water). We can write the molar volume of water as

$$V = x_{HB}\overline{V}_{HB} + (1 - x_{HB})\overline{V}_{NHB} \qquad (10.39)$$

where $\overline{V}_{HB}$ and $\overline{V}_{NHB}$ are the partial molar volumes of the *HB*'ed and the non-*HB*'ed species, respectively. The temperature dependence of the volume is

$$\left(\frac{\partial V}{\partial T}\right)_P = x_{HB}\frac{\partial \overline{V}_{HB}}{\partial T} + (1 - x_{HB})\frac{\partial \overline{V}_{NHB}}{\partial T}$$

$$+ (\overline{V}_{HB} - \overline{V}_{NHB})\left(\frac{\partial x_{HB}}{\partial T}\right)_P \qquad (10.40)$$

The first two terms on the rhs of the equation are presumed to be positive (i.e. the two- component mixture is presumed to behave "normally." Since $(\overline{V}_{HB} - \overline{V}_{NHB})$ is presumed to be positive (due to the relatively more open structure of the *HB*'ed species), we can expect a large negative contribution from the third term on the rhs of Eq. (10.40). In order to get a net negative temperature dependence of the molar volume, we need to get a large negative contribution from the third term on the rhs of Eq. (10.40).

10.2 A Cluster Model for Aqueous Solutions of Inert Molecules

In this section, we will discuss one aspect of aqueous solutions which has been studied for over 70 years. The experimental fact is the following: The solvation entropy of inert solutes in water is large and negative, much larger than the solvation entropy of the same molecule in any other liquid. There are many outstanding properties of aqueous solutions, but the entropy of solvation is perhaps one of the most outstanding property that has captured the imagination of many scientists for a very long period of time.

The experimental fact is that the entropy of solvation of say, argon is outstandingly large and negative compared with the corresponding quantity in a series of alcohols. The question is why? What are the molecular reasons for such an outstanding quantity?

In 1945 Frank and Evans speculated that when a non-polar molecule such as argon or methane enters water, it creates some kind of *iceberg*. This seminal idea is still being quoted today and is considered to be a major breakthrough in our understanding of the properties of aqueous solutions. A closer examination of the iceberg formation idea leads inevitably to the conclusion that this is not much more than an equivalent statement of the experimental facts.

If entropy is considered to be a measure of disorder, then negative change in entropy is equivalent to an increase in order. More *structure* is considered to be more order. Therefore, the statement of iceberg formation is tantamount to the statement of more structure, hence more order, hence negative entropy change. All these are equivalent statements of the experimental facts. None of the statements *explains why* an inert solute such as argon would create icebergs, increase order, increase the structure, or decrease entropy.

In 1969, Lovett and Ben-Naim published an article on "one-dimensional model for aqueous solutions of inert gases". The main objective of that paper was to show how clusters of water-like particles could be promoted by inert solute. This phenomenon was supposed to mimic the enhancement of the structure of water induced by inert solutes, and thereby provides an explanation of the outstanding large and negative entropy and enthalpy of solvation.

10.2.1 Application of the Cluster Model for Water to the Study of Dilute Solutions of Inert Solutes

In this section, we extend the application of the cluster model discussed in Section 10.1.3, to examine the solvation thermodynamics of simple solutes in the water-like solvent. The model is schematically described in Fig. 10.12. The only new feature that is added here, compared with the model discussed in Section 10.1.3, is that the clusters of water-like

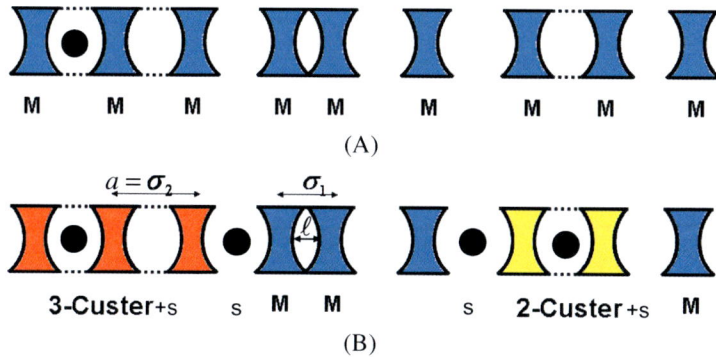

Fig. 10.12 The primitive model, (A), and the cluster model for aqueous solutions of an inert solute S, (B).

particles contain "holes" in which solute molecules can be accommodated. This feature is the analog of the cavities formed by the network of hydrogen bonded molecules in real water.

In this section we show that one can obtain both large solvation Gibbs energy (or low solubility) and large negative entropy and enthalpy of solvation, in the same model by choosing a larger distance for the *HB* energy.

In the 1D cluster model we assume that there are $(n - 1)$ holes (or cavities) per n-cluster, and a hydrogen bonding distance of $\sigma_2 = 2$ (where $\sigma_1 = 1$ was the hard core diameter). Clearly, one can cause a *decrease* in the solubility of the solute either by decreasing the number of holes or by increasing the *HB* distance (i.e. making the size of a cluster of *HB*'ed particles larger). In the next section we present some technical details on the PF of the system. The reader who is not interested in those details can skip this section and continue to Section 10.2.3.

10.2.2 (***) *The Partition Function and the Computational Procedure*

The partition function for the system of two-component, water-like and inert solute particles is fairly complicated. Here, we present the final form of the partition function and the computational procedure to obtain the cluster-size distribution, and the solvation quantities in very dilute solution.

Extending the same procedure as described in Section 10.1, we can write the Laplace transform of the grand partition function as

$$\psi(\Xi(T,L,\lambda)) = \int \exp[-\beta PL]\Xi(T,L,\lambda)dL$$

$$= \sum_{N=0}^{\infty} Tr(M^N) \qquad (10.41)$$

where M is a matrix with the elements

$$M_{1,0} = \frac{\lambda_0}{\Lambda_0}\psi_{SS}, \quad M_{0,1} = M_{1,0} = \left(\frac{\lambda_0\lambda_1}{\Lambda_0\Lambda_1}\right)^{1/2}\psi_{SW}$$

$$M_{0,ni} = M_{ni,0}$$

$$= \left[\frac{\lambda_{ni}\lambda_0}{\Lambda_1^n\Lambda_0^{i+1}}\psi_{HB}^{(n-1)}\exp[-\beta\varepsilon_{ni}]l^i\binom{n-1}{i}\right]^{1/2}\psi_{SW}$$

$$M_{1,1} = \frac{\lambda_1}{\Lambda_1}\psi_{ww},$$

$$M_{1,ni} = M_{ni,1}$$

$$= \left[\frac{\lambda_{ni}\lambda_1}{\Lambda_1^{n+1}\Lambda_0^i}\psi_{HB}^{(n-1)}\exp[-\beta\varepsilon_{ni}]l^i\binom{n-1}{i}\right]^{1/2}\psi_{WW}$$

$$M_{ni,mj} = M_{mj,ni}$$

$$= \left[\frac{\lambda_{ni}\lambda_{mj}}{\Lambda_1^{m+n}\Lambda_0^{i+j}}\psi_{HB}^{m+n-2}\exp[-\beta(\varepsilon_{ni}+\varepsilon_{mj})]\right.$$

$$\left.\times l^{(i+j)}\binom{n-1}{i}\binom{m-1}{j}\right]^{1/2}\psi_{WW} \qquad (10.42)$$

where

$$\varepsilon_{ni} = (n-1)\varepsilon_{HB} + i\varepsilon_0 \qquad (10.43)$$

ε_0 is the interaction energy of the solute with the hole in the cluster. ε_{HB} is the *HB* energy, l is the "free volume", which is essentially the size of the holes in the cluster. ψ_{HB} is defined by

$$\psi_{HB} = \int_0^\infty dR \exp\left[-\beta PR - \beta\frac{1}{2}k(R-a)^2\right] \quad (10.44)$$

where k is a force constant for the harmonic potential function between two particles engaged in an *HB*, and a is the *HB* distance.

It is convenient to transform the elements of the matrix M by defining the quantities

$$t_0^2 = \frac{\lambda_0}{\Lambda_0}\left[\frac{\psi_{SW}}{\psi_{WW}}\right]^2, \quad t_1^2 = \frac{\lambda_1}{\Lambda_1} = \lambda_1 q_1$$

$$t_{ni}^2 = \frac{\lambda_{ni}}{\Lambda_1^n \Lambda_0^i}\psi_{HB}^{(n-1)}\exp[-\beta\varepsilon_{ni}]l^i\binom{n-1}{i}$$

$$= \lambda_{ni}q_{ni} \quad (10.45)$$

With Eq. (10.45) one can transform the matrix M into a new matrix M' having the same eigenvalues, but a simpler form

$$M' = \psi_{WW}\begin{pmatrix} \alpha t_0^2 & tt_0 & 0 & 0 & 0 & \cdots \\ tt_0 & t^2 & 0 & 0 & 0 & \cdots \\ 0 & 0 & 0 & 0 & 0 \\ \vdots & & & & & \end{pmatrix} \quad (10.46)$$

where $\alpha = \psi_{WW}\psi_{SS}/\psi_{SW}^2$, and

$$t^2 = t_1^2 + \sum_{n=2}^\infty \sum_{i=0}^{n-1} t_{ni}^2 \quad (10.47)$$

Now, it is easy to find the trace of the matrix M' which is the same as the trace of the matrix M. The two eigenvalues are

$$\gamma_\pm = \frac{\psi_{WW}}{2}\{\alpha t_0^2 + t^2 \pm [(\alpha t_0^2 - t^2)^2 + 4t^2 t_0^2]^{1/2}\}$$

(10.48)

The correct solution is the one with the plus sign (for $t_0^2 = 0$, we recover the equation for pure water). Using the eigenvalue γ_+ in (10.48) in the Laplace transform of the grand partition function, and requiring that this "generalized" partition function diverges, one gets the equation of state of this system in the form

$$F(T, P, \lambda) = \frac{\psi_{WW}}{2}\{\alpha t_0^2 + t^2 + [(\alpha t_0^2 - t^2)^2 + 4t^2 t_0^2]^{1/2}\}$$
$$= 1$$

(10.49)

We use the equation of state $F(T, P, \lambda) = 1$ for all the following computations. Note that in this equation, all the λ_{ni} are independent variables, i.e. the system is viewed as a mixture of clusters characterized by n monomers and accommodating i solute particles.

In all the calculations carried out for this system, we first treat the absolute activities λ_{ni} as independent variables. At the final expression, we must evaluate the result at equilibrium, i.e. we must introduce the condition $\mu_{ni} = n\mu_i + i\mu_0$ or equivalently

$$\lambda_{ni} = \lambda_1^n \lambda_0^i$$

(10.50)

From (10.49), we can express t^2 in terms of t_0^2

$$t^2 = \frac{\alpha \, t_0^2 \psi_{WW} - 1}{\psi_{WW}[\alpha \, t_0^2 \psi_{WW} - \psi_{WW} t_0^2 - 1]} \tag{10.51}$$

Note that for $t_0^2 = 0$, we recover the equation of state for pure water.

$$t^2 \psi_{WW} = 1 \tag{10.52}$$

The number densities of the various species are calculated from

$$\rho_{ni} = \frac{N_{ni}}{\langle L \rangle} = -\frac{\partial F / \partial \mu_{ni}}{\partial F / \partial P} = -\frac{(\partial F / \partial \lambda_{ni}) \, \beta \lambda_{ni}}{\partial F / \partial P} \tag{10.53}$$

and

$$\rho_i = \frac{-(\partial F / \partial \lambda_i) \beta \lambda_i}{\partial F / \partial P} \tag{10.54}$$

$$\rho_0 = \frac{-(\partial F / \partial \lambda_0) \beta \lambda_0}{\partial F / \partial P} \tag{10.55}$$

All the derivatives between (10.53), (10.54), and (10.55) can be obtained from the equation of state (10.49). We are interested only in ratios of the densities, or mole fractions. These are

$$\frac{\rho_{ni}}{\rho_1} = \frac{\lambda_{ni}(\partial F / \partial \lambda_{ni})}{\lambda_i (\partial F / \partial \lambda_i)} = \frac{t_{ni}^2}{t_1^2} \tag{10.56}$$

$$\frac{\rho_0}{\rho_1} = \frac{\lambda_0(\partial F / \partial \lambda_0)}{\lambda_i(\partial F / \partial \lambda_i)} = \frac{t_0^2 \left[\alpha + \frac{4t^2 + 2\alpha(\alpha t_0^2 - t^2)}{\sqrt{}} \right]}{t_1^2 \left[1 + \frac{t^2 + 2t_0^2 - \alpha t_0^2}{\sqrt{}} \right]} \tag{10.57}$$

where

$$\sqrt{\ } = [4t^2 t_0^2 + (t^2 - \alpha t_0^2)^2]^{1/2} \tag{10.58}$$

The total number density of water molecules is

$$\rho_W = \rho_1 + \sum_{n,i} n\rho_{ni} = \rho_1 \left[1 + \sum_{n,i} n\frac{\rho_{ni}}{\rho_1} \right]$$

$$= \rho_1 \left[1 + \sum_{n,i} n\frac{t_{ni}^2}{t_1^2} \right] \tag{10.59}$$

and the total number density of solute molecules is

$$\rho_S = \rho_0 + \sum_{n,i} i\rho_{ni} = \rho_1 \left[\frac{\rho_0}{\rho_1} + \sum_{n,i} i\frac{\rho_{ni}}{\rho_1} \right]$$

$$= \rho_1 \left[\frac{\rho_0}{\rho_1} + \sum_{n,i} i\frac{t_{ni}^2}{t_1^2} \right] \tag{10.60}$$

Next, we define the mole fraction of clusters of the size n

$$X_n = \frac{\sum_{i=0}^{n-1} \rho_{ni}}{\rho_W} \tag{10.61}$$

and the number of hydrogen bonds (HB) per water molecule as

$$X_{HB} = \sum_{n=2}^{\infty} (n-1)X_n \tag{10.62}$$

The mole fraction of the solute S in the system is

$$Y_s = \frac{\rho_s}{\rho_s + \rho_W} = \frac{1}{1 + \left(\frac{\rho_W}{\rho_s}\right)} \tag{10.63}$$

For the numerical calculations of X_n and X_{HB}, we proceed as follows.

First, we can evaluate the sum

$$t^2 = t_1^2 + \sum_{n=2}^{\infty}\sum_{i=0}^{n-1} t_{ni}^2 = \lambda_1 q_1 + \sum_{n=2}^{\infty}\sum_{i=0}^{n-1} \lambda_{ni} q_{ni}$$

$$= \frac{t_1^2}{1 - \eta t_1^2(1 + \xi t_0^2)} \tag{10.64}$$

where q_{ni} are defined in (10.45), and where

$$\eta = \psi_{HB}\exp[-\beta\varepsilon_{HB}] \quad \text{and} \quad \xi = \left[\frac{\psi_{WW}}{\psi_{WS}}\right]^2 l \exp[-\beta\varepsilon_0]$$

The last relation can be inverted to obtain

$$t_1^2 = \frac{t^2}{1 + \eta t^2(1 + \xi t_0^2)} \tag{10.65}$$

Since t^2 is already given as a function of T, P, λ_S, in (10.51), we have also t_1^2 as a function of T, P, λ_S (note that λ_s is the same as λ_0 at equilibrium).

The average number of HBs per molecule is given by

$$X_{HB} = \sum_{n=2}^{\infty}(n-1)X_n$$

$$= \sum_{n=2}^{\infty} n X_n - \sum X_n = 1 - \sum X_n$$

$$X_{HB} = \eta t_1^2 (1 + \xi t_0^2) \qquad (10.66)$$

The two quantities X_n and X_{HB} are related by

$$X_n = \frac{\rho_n}{\rho_W} = \frac{\sum_{i=0}^{n-1} \rho_{ni}}{\rho_W} = \frac{\sum_{i=0}^{n-1} t_{ni}^2 / t_1^2}{1 + \sum_{i=0}^{n-1} n t_{ni}^2 / t_1^2}$$

$$= X_{HB}^{n-1} (1 - X_{HB})^2 \qquad (10.67)$$

This is all we need to calculate the distribution of species and the average number of HBs, sometimes referred to as the *structure* of the system. In real water, a larger number of HBs in the system is indeed associated with a higher degree of structure. In the 1D model, it is difficult to associate the concept of structure (as well as of order) with the number of HB.

Next, we turn to the calculation of the solvation quantities. We are interested only in the limit of infinite dilution system hence, we expand the equation of state to first order in t_0^2 to obtain

$$\psi_{WW} (t^2 - t_0^2) = 1 \qquad (10.68)$$

Since t_0^2 is proportional to λ_0, and t^2 is expressed in terms of t_0^2, we can extract λ_0, or equivalently the chemical potential of the solute to obtain

$$\mu_S = (T, P, \rho_S) = T \ln \rho_S \Lambda_0$$

$$+ T \ln \left[\frac{\exp[\beta \varepsilon_0] (\psi_{ii} + \exp[\beta \varepsilon_{HB}] \psi_{WW})}{\rho_W [l \psi_{HB} + \exp[\beta (\varepsilon_0 + \varepsilon_{HB})] \psi_{SW}^2} \right]$$

$$(10.69)$$

From (10.69) we get the solvation Gibbs energy at infinite dilution

$$\Delta\mu_S^* = \mu_S(T, P, \rho_S) - T \ln \rho_S \Lambda_0$$

$$= T \ln \left[\frac{\exp[\beta\varepsilon_0](\psi_{HBi} + \exp[\beta\varepsilon_{HB}]\psi_{WW}}{\rho_W[l\psi_{HBi} + \exp[\beta(\varepsilon_0 + \varepsilon_{HB})]\psi_{WW}^2} \right]$$

$$(10.70)$$

10.2.3 Results for Hard-Rod Solutes in Dilute Solutions in the Primitive Cluster Model

Figure 10.13 shows the dependence of X_1 (the mole fraction of the monomeric "water" particles) as a function of the mole fraction of the solute y_S. The behavior of x_1 as a function of y_S is unique in the sense that initially at low values of y_S, X_1 decreases with y_S. This means that monomers are used to build-up clusters. As we will see below, this behavior is typical to aqueous solutions. It is part of the molecular explanation of the large negative entropy and enthalpy of solvation of the solute. At higher concentrations of the solute S, clusters must

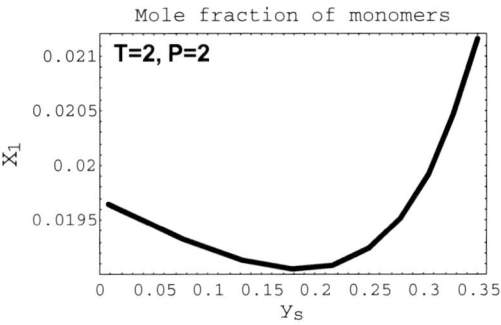

Fig. 10.13   The mole fraction X_1 as a function of the mole fraction of the solute S.

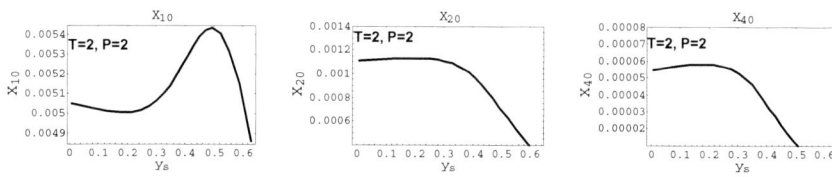

Fig. 10.14 The mole fractions X_2, X_{20}, and X_{40} as a function of the mole fraction of the solute S.

be disintegrated and eventually water monomers become diluted in S.

For some intermediated values of n (Fig. 10.14), there is an initial values of n, decrease of X_n as we add S, then X_n increases but eventually decreases to zero as y_S increases to unity. For larger clusters, say $n \geq 20$, the mole fraction X_i initially increases, but eventually decreases by concentration of the solute. The details of which cluster size is stabilized or destabilized by the solute depends on the temperature and pressure, but the typical behavior of X_1 and X_n for small n, and X_n for large n is common in the entire range of temperatures and pressures we are interested in.

To understand the outstanding large and negative entropy and enthalpy of solvation, it is not necessary to examine the detailed variation of each X_n. Instead, we examine the average number of HBs per water molecules, which we denote by $\langle XHB \rangle$. This is a measure of the overall "structure" of the system. The variation of this quantity with y_s determines the value of the entropy and the enthalpy of solvation of the solute s.

Figure 10.15 shows the dependence of $\langle XHB \rangle$ on y_S for $T = 2$ and $P = 2$. It is seen that initially $\langle XHB \rangle$ increases with y_S. Beyond certain value of y_S, it starts to decrease towards zero, when water becomes diluted in S.

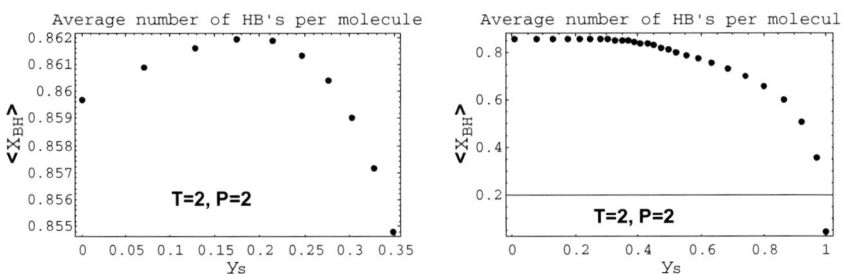

Fig. 10.15   Dependence of $\langle XHB \rangle$ on the mole fraction of solute concentration of S at low and high concentrations of S.

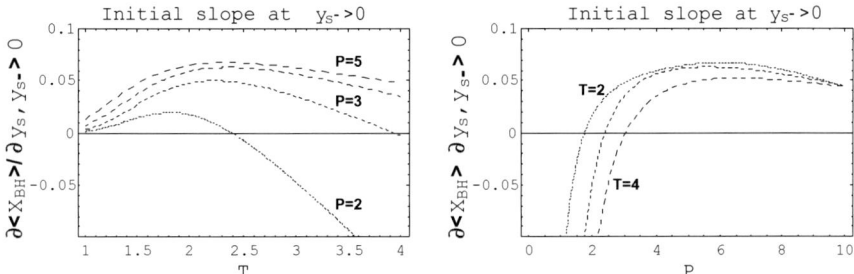

Fig. 10.16   Dependence of the limiting slope of $\langle XHB \rangle$ at $y_S \to 0$ on the temperature and the pressure.

We are interested here only in the solvation properties in the limit of $y_S \to 0$, only the slope of $\langle XHB \rangle$ at $y_S \to 0$ is of interest. Figure 10.16 shows the limiting slope

$$\lim_{y_s \to 0} \frac{\partial \langle XHB \rangle}{\partial y_S} \tag{10.71}$$

as a function of T and P. The larger the limiting slope, the larger the "stabilization of the structure", or larger the enhancement of the formation of HB's in the system. Clearly, cluster formation requires energy. This formation of HB will contribute large and negative enthalpy and energy of

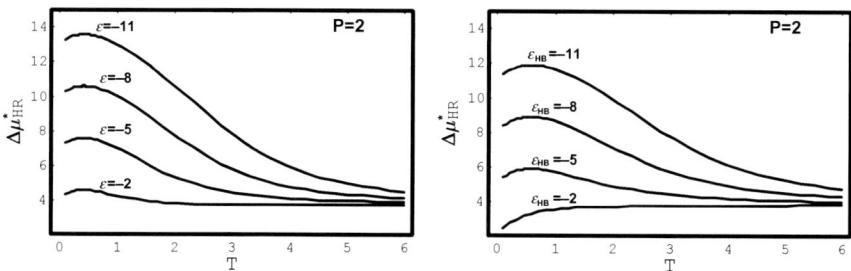

Fig. 10.17 Solvation Gibbs energy of *HR* in *HR* and in *HB* solvents for various values of the *HB* energy ε_{HB}.

solvation. The same argument holds for the large negative entropy of solvation.

Figure 10.17 shows the solvation Gibbs energies of an *HR* in the "water" (*HB*) and in the *HR* for different values of ε_{HB} in the former, and ε in the latter. The values of $\Delta\mu_S^*$ are large and positive, corresponding to low solubility compared with the solubility in the *HR* solvent, as well as in the square-well solvent. Also, the maximum of $\Delta\mu_S^*$ as a function of *T* is in accordance with the experimental data on solvation of inert gases in water. Note that as $|\varepsilon_{HB}|$ increases, we get larger and larger $\Delta\mu_S^*$. Note also that the slope of $\Delta\mu_S^*$ versus *T* is large and positive, but not very sensitive to variation of ε_{HB} in the range of temperatures between 1 and 3. This corresponds to the entropy of solvation discussed below.

Figure 10.18 shows the values of ΔS_S^* for an *HR* in the cluster model with two values of $\varepsilon_{HB} = -10, -12$. We also show ΔS_S^* in the *HR*. The difference is quite striking. The solvation entropy is about ten times larger in the cluster model than in the *HR* as well as in the square-well model. In this model, ΔS_S^* values also cross at about $T = 1.5$. (It crosses again at $T \approx 3$, but this is not the range of temperatures of

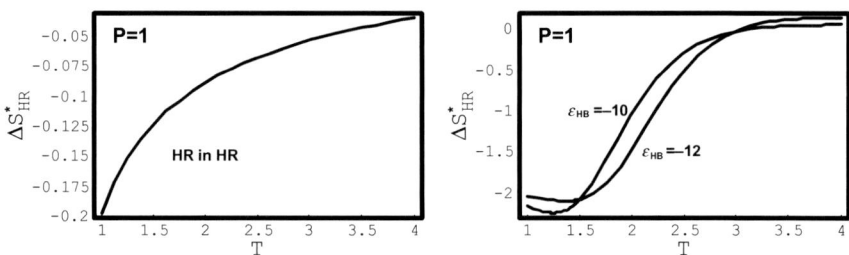

Fig. 10.18 The solvation entropy of *HR* in *HR* and in the cluster model (*HB*).

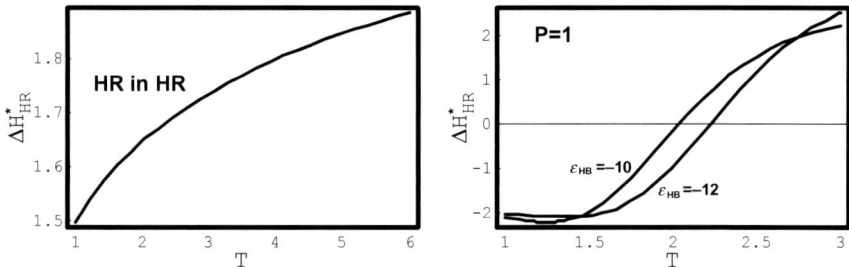

Fig. 10.19 The solvation enthalpy of *HR* in *HR* and in the cluster model (*HB*).

interest in this study). Below $T = 1.5$, the values of ΔS_S^* are more negative for the "H_2O" than for the "D_2O". This is in accordance with the isotope effect on the entropy of solvation of neon, krypton and xenon in real H_2O and D_2O, but not for argon for which ΔS_S^* is more negative in D_2O.

Figure 10.19 shows the values of the solvation enthalpies in the cluster model and in the square-well solvent. Again, we find that while ΔH_S^* in the square-well solvent are *positive*, they are large and negative in the cluster model. The isotope effect is relatively small and as in the case of the entropy, some gases have a positive isotope effect on ΔH_S^*, and some negative.

10.2.4 *Some Concluding Remarks*

It is remarkable that simple 1D models can exhibit some of the outstanding properties of water, which is considered to be one of the most complicated liquids. The success of the primitive model in mimicking the properties of pure water confirms the conjecture that the most important aspect of the molecular interactions in water is that they can produce a correlation between low local density and large binding energy. At first sight, one may think that the success of the model to mimic some outstanding properties of aqueous solutions is even more surprising. However, there is a deep connection between the sources of the anomalous behavior of pure liquid water, and the outstanding values of the entropy and enthalpy of solvation of inert solutes in water. Both phenomena are intimately connected, and both originate from the same characteristic molecular interactions in water. The large negative entropy and the enthalpy of solvation of an inert solute (even a hard sphere solute), is a result of the "stabilization" of the "structure" of water. This, when translated into the language of the 1D model means that the inert solute enhances the formation of HB'ed cluster. This enhancement requires energy; hence, the large negative enthalpy (and energy) of solvation. The entropy of solvation is usually interpreted by invoking the concept of *structure* or *order* in the solvent. However, the exact entropy–enthalpy compensation principle states that whatever the contribution to ΔH_S^* due to structural changes in the solvent is, the same is true for the contribution to the entropy of solvation. In the most general formulation of this principle, we can write the expressions for $\Delta \mu_S^*$, ΔS_S^* and

ΔE_S^* as[1]

$$\Delta \mu_S^* = -k_B T \langle \exp(-\beta B_S) \rangle_0 \tag{10.72}$$

$$\Delta S_S^* = k_B \ln \langle \exp(-\beta B_S) \rangle_0$$
$$+ \frac{1}{T}[\langle B_S \rangle_0 + \langle U_N \rangle_S - \langle U_N \rangle_0] \tag{10.73}$$

$$\Delta E_S^* = \langle B_S \rangle_0 + \langle U_N \rangle_S - \langle U_N \rangle_0 \tag{10.74}$$

Here, B_S is the total interaction energy of the solute s with all solvent molecules being at a fixed configuration $R_1, R_2, \ldots, R_N$. The symbol $\langle \ \rangle_0$ indicates an average over all the configurations of the solvent molecules before the insertion of S. The quantity $\langle U_N \rangle_S - \langle U_N \rangle_0$ is simply the change in the total interaction energy among all solvent molecules induced by the insertion of the solute s at some fixed position in the system. This term was originally referred to as "structural" changes in the solvent. The important point is that this term is a change in *energy*, and the same terms appear both in (10.73) and (10.74). It just happened that in water, the more "structured" arrangement coincides with the larger value of $|\langle U_N \rangle|$. However, the *structure*, in itself, is not essential to understanding the large negative value of the entropy of solvation. It is the shift, induced by s, towards stronger interaction energy among the solvent molecule that explains ΔE_S^*, as well as part of ΔS_S^*. The choice of an *HR*, as a solute, makes $\langle B_S \rangle_0$ equals zero. Hence, all of ΔE_S^* is due to the difference $\langle U_N \rangle_S - \langle U_N \rangle_0$ which, may be referred to as structural change induced by the solute on the solvent. The *same* quantity divided by T also appears in ΔS_S^* and this very term also consists of the explanation of the large negative entropy of solvation.

As one can see from (10.72), the solvation Gibbs energy is independent of the structural changes induced in the solvent. This is exactly the reason why we could have seen the anomalous large positive Gibbs energy of solvation in both the primitive and the cluster models, in the 1D model, the solubility of the HR depends only on the total interaction energy action among the solvent molecules and not on the structural changes induced by the solute. On the other hand, it is the structural changes induced by the solute that explains both ΔE_S^* and ΔS_S^* in water. That is also the reason why we could observe large negative values of ΔE_S^* (or ΔH_S^*) and ΔS_S^* in the cluster model but not in the primitive model.

References

References to specific articles can be found in the following books:

A. Ben-Naim. (2001) *Cooperativity and Regulation in Biochemical Systems*. Plenum Press, New York.

A. Ben-Naim. (2006) *Molecular Theory of Solutions*. Oxford University Press, Oxford.

A. Ben-Naim. (2007) *Entropy Demystified*. World Scientific, Singapore.

A. Ben-Naim. (2008) *A Farewell to Entropy, Statistical Thermodynamics Based on Information*. World Scientific, Singapore.

A. Ben-Naim. (2009) *Molecular Theory of Water and Aqueous Solutions, Part I: Understanding Water*. World Scientific, Singapore.

A. Ben-Naim. (2010) *Molecular Theory of Water and Aqueous Solutions Part II: The Role of Water in Protein Folding, Protein Association and Molecular Recognition*. World Scientific, Singapore.

A. Ben-Naim. (2012) *Entropy and the Second Law: Interpretation and Miss-Interpretationsss*. World Scientific, Singapore.

Notes

Notes to Chapter 1

1. For a more detailed discussion on this refer to Ben-Naim (2008, 2009, 2010, 2011).
2. For more details, see Ben-Naim (1992, 2008).
3. For details, see Ben-Naim (2008).
4. For details, see Ben-Naim (1992, 2006).

Notes to Chapter 2

1. Shannon (1948).
2. For more details, see Ben-Naim (2008, 2012).
3. We will always use this function for $0 \leq p \leq 1$. Note that $0 \log 0 = 0$. This can be shown by taking the limit $\lim_{p \to 1} p \log p$. See Section 2.5.
4. Note that "*bit*" is used here as a unit of information. It is derived from *binary digit*. However, the latter is also used for either 0 or 1.
5. For more details, see Ben-Naim (2008, 2012).

6. We neglect the possibility that the dart hit any *line* bordering two areas.
7. For more details, see Ben-Naim (2008).
8. For details, see Ben-Naim (2008).
9. See Ben-Naim (2008) and (2012).
10. See Ben-Naim (2008).

Notes to Chapter 3

1. For more details, see Ben-Naim (2008, 2010, 2012).
2. For more details, see Ben-Naim (2008, 2012).
3. For details, see Ben-Naim (2008).
4. For details, see Ben-Naim (2012).
5. For more details, see Ben-Naim (2012).
6. For some examples, see Ben-Naim (2013).

Note to Chapter 5

1. For details, see Ben-Naim (2001).

Notes to Chapter 6

1. For more details, see Ben-Naim (2001).
2. For more details, see Ben-Naim (2001).
3. For more details, see Ben-Naim (2001).
4. For more details, see Ben-Naim (2001).
5. For details, see Ben-Naim (2001).
6. More details may be found in Ben-Naim (2001).
7. Further discussion of these systems may be found in Ben-Naim (2001).

Notes to Chapter 7

1. For other possibilities of assigning the factors to particles, see Ben-Naim (1992).
2. See Ben-Naim (1992).

Notes to Chapter 9

1. For more details, see Ben-Naim (1992, 2006).
2. For more details, see Ben-Naim (2006).
3. The pair correlation function in the theory of solutions is defined in the same manner as we have defined the correlations between two events in Chapter 6.
4. For more details, see Ben-Naim and Santos (2009).
5. For details, see Ben-Naim (2006).
6. For details, see Ben-Naim (2006).

Notes to Chapter 10

1. For more details, see Ben-Naim (2009).
2. Ben-Naim (1992).
3. Lovett and Ben-Naim (1969).

Index

Made in the USA
Monee, IL
12 February 2020